Studies of the Americas

Series Editor
Maxine Molyneux, Institute of the Americas, University College London, London, UK

The Studies of the Americas Series includes country specific, cross-disciplinary and comparative research on the United States, Latin America, the Caribbean, and Canada, particularly in the areas of Politics, Economics, History, Sociology, Anthropology, Development, Gender, Social Policy and the Environment. The series publishes monographs, readers on specific themes and also welcomes proposals for edited collections, that allow exploration of a topic from several different disciplinary angles.

This series is published in conjunction with University College.

London's Institute of the Americas under the editorship of Professor Maxine Molyneux.

More information about this series at
https://link.springer.com/bookseries/14462

Cecilia T. Lanata-Briones · Andrés Estefane ·
Claudia Jorgelina Daniel
Editors

Socio-political Histories of Latin American Statistics

palgrave
macmillan

Editors
Cecilia T. Lanata-Briones
Department of Economics
University of Warwick
Coventry, UK

Andrés Estefane
Independent Researcher
Santiago, Chile

Claudia Jorgelina Daniel
Centro de Investigaciones Sociales
Consejo Nacional de Investigaciones
Científicas y Técnicas
Buenos Aires, Argentina

Studies of the Americas
ISBN 978-3-030-87716-3 ISBN 978-3-030-87714-9 (eBook)
https://doi.org/10.1007/978-3-030-87714-9

This Palgrave Macmillan imprint is published by the registered company Springer Nature Switzerland AG
The registered company address is: Gewerbestrasse 11, 6330 Cham, Switzerland

Contents

Notes on Contributors

Alexandre de Paiva Rio Camargo is Associate Professor at the University Research Institute of Rio de Janeiro (IUPERJ) of the Candido Mendes University (UCAM, Brazil) and former Visiting Professor at the School of Advanced Studies in the Social Sciences (EHESS, France). He is a political and historical sociologist, with a research agenda focused on the politics of numbers, vital and ethno-racial statistics, the role of quantification in the constitution of scientific disciplines, and nationalism and state building, with a regional focus on Brazil. He is the author of numerous book chapters and journal articles.

Laura Cházaro García is full-time Professor-Researcher in the Departamento de Investigaciones Educativas del Centro de Investigación y de Estudios Avanzados del Instituto Politécnico Nacional (Mexico). Her main research interests include history of sciences, nineteenth-century Mexico, social history of medical and statistical thinking, quantification, and bodies measurements.

Claudia Jorgelina Daniel holds a Ph.D. in Social Sciences from the University of Buenos Aires (UBA, Argentina). Currently, she is Adjunct Researcher at the Consejo Nacional de Investigaciones Científicas y Técnicas based at the Centro de Investigaciones Sociales (CIS-CONICET/IDES, Argentina) and Lecturer at the School of Social Sciences (UBA, Argentina). Her research focuses on the study of the Argentine institutions that produce official statistics from a socio-historical perspective. Her interests range from the production to the social and political uses of statistics in contemporary societies. She is the author of *Números Públicos. Las estadísticas en Argentina (1990–2010)* (Fondo de Cultura Económica, 2013). She is a founding member of the Asociación de las Américas por la Historia de las Estadísticas y el Cálculo de las Probabilidades (AAHECP).

Andrés Estefane holds a Ph.D. in History from the State University of New York at Stony Brook, and is an Independent Researcher based in Santiago, Chile. His research explores the links among bureaucracy, local power, and the production of state knowledge in historical perspective. He co-edited the four-volume project *Historia política de Chile, 1810–2010* (Fondo de Cultura Económica, 2017–2018) and recently published *Cuando íbamos a ser libres*, an annotated edition of primary sources on the history of liberalism in Chile (Fondo de Cultura Económica, 2021).

Victoria Estrada Orrego holds a Ph.D. in History from the School of Advanced Studies in the Social Sciences (EHESS, France). She is Lecturer at the Metropolitan Technological Institute (Medellin, Colombia). Her work examines the history of medicine and statistics in Colombia during the last decades of the nineteenth century and the first half of the twentieth century.

Cecilia T. Lanata-Briones is Assistant Professor in the Department of Economics of the University of Warwick (UK) and Adjunct Researcher of the Centro Interdisciplinario para el Estudio de Políticas Públicas (Ciepp, Argentina). She holds a Ph.D. in Economic History from the London School of Economics and Political Science (UK). Her research interests focus on the production and use of Latin American economic statistics from a socio-historical perspective.

Hernán Otero is a Historian and Demographer. He holds a Ph.D. from the School of Advanced Studies in the Social Sciences (EHESS, France). He is Principal Researcher at the Consejo Nacional de Investigaciones Científicas y Técnicas (CONICET, Argentina) and teaches sociology of population at the Universidad Nacional del Centro de la Provincia de Buenos Aires (Tandil, Argentina). He is a member of the Academia Nacional de Historia de la República Argentina. His areas of research concern the history of population, especially international migration, the formation of the Argentine statistical system, and, more recently, the history of the elderly.

Marcelo Somarriva holds a Ph.D. in History from University College of London (UK). He is Assistant Professor at the Faculty of Liberal Arts, Universidad Adolfo Ibáñez (Chile). His research focuses on the relationship between Britain and South America, with emphasis on the cultural encounters during the Enlightenment. He is currently conducting research on the influence of climate theory and the political language of "public happiness" in the construction of the colonial imaginary in Latin America.

Abbreviations

DGE	Dirección General de Estadísticas (General Board of Statistics), Mexico
DGE-B	Diretoria Geral de Estatística (General Board of Statistics), Brazil
DNE	Departamento de la Estadística Nacional (Department of National Statistics)
IBGE	Instituto Brasileiro de Geografia e Estatística (Brazilian Institute of Geography and Statistics)
IHGB	Instituto Histórico e Geográfico Brasileiro (Brazilian Historical and Geographical Institute)
NAM	Academia Nacional de Medicina (National Academy of Medicine)
SHC	Consejo Superior de Salubridad (Superior Health Council)
SMGyE	Sociedad Mexicana de Geografía y Estadística (Mexican Society for Geography and Statistics)

List of Figures

List of Tables

CHAPTER 1

Introduction

Claudia Jorgelina Daniel, Cecilia T. Lanata-Briones, and Andrés Estefane

This book brings together works representative of the themes and approaches that have informed the studies on the socio-political history of Latin American statistics after independence. These investigations illustrate the robustness and future perspectives of an area of research that has shown a remarkable dynamism in recent decades. This publication, however, not only assembles selected works on and from the region. It also aims to build bridges between Latin American investigations and

C. J. Daniel
Centro de Investigaciones Sociales, Consejo Nacional de Investigaciones Científicas y Técnicas, Buenos Aires, Argentina
e-mail: cdaniel@ides.org.ar

C. T. Lanata-Briones (✉)
Department of Economics, University of Warwick, Coventry, UK
e-mail: cecilia.lanata-briones@warwick.ac.uk

A. Estefane
Independent Researcher, Santiago, Chile

C. T. Lanata-Briones et al. (eds.), *Socio-political Histories of Latin American Statistics*, Studies of the Americas,
https://doi.org/10.1007/978-3-030-87714-9_1

those that explore the trajectory of statistics at an international level. Given that only a small percentage of the research on the history of Latin American statistics is published in English, the trajectory of the region tends to have a marginal or dispersed presence in global and transnational approaches. By bringing Latin American academic production closer to wider audiences, this book seeks to foster substantive dialogues that could enrich the understanding of the socio-political history of this science from a truly global perspective.

The studies published in this volume were guided by diverse questions, usually defined by the national experiences they focused on. However, they converge in some common concerns, which account for the type of questions the socio-political history of statistics has been exploring since the end of the twentieth century. One of those concerns focuses on the emphasis on the social and political factors that make possible and determine the production of official numbers. These are works that examine very closely both the institutions that produce statistics and the modes of production of these figures, while always considering the structural conditions in which those processes of knowledge production took place. Likewise, the chapters of this book follow a similar approach when scrutinising the problem of legitimacy and/or authority of any enumeration or counting operation, particularly of those led by the state. While neither legitimacy nor authority are taken for granted, the contributors to this book carefully explore the efforts to claim objectivity, universality, or even usefulness of the statistical knowledge on the part of the historical agents identified with this science. In doing so, these works embody what French sociologist Alain Desrosières aspired to when thinking of the need to reassess "statistical reasoning as a mode of abstraction into a more general social or political history" (1998, 324).

In more general terms, these studies are samples of an intellectual effort aimed to reconstruct the institutional, cultural, and political conditions of the production of statistical knowledge, taking also into account the objectification work carried out by statisticians. This task implies taking distance from the understanding of statistics as a phenomenon detached from producers and users of numbers or as unproblematic and self-evident facts. Following Desrosières (1998), statistics create new objects by combining different and heterogeneous elements which in turn follow a common principle of equivalence. A socio-political historical approach to statistics is characterised by the effort to grasp the meaning of those creations, restoring the structure of consensus—and of controversy—that makes statistical measurements hold as stable social and political artefacts at a given historical moment.

Such an approach traces the conventional aspects of statistical practices to identify how they become naturalised norms and standards as well as technical requirements. Following it, scholarsstudy the official production of statistics without losing sight of the social, cultural, and political factors that influence the contents and limitations of statistical research programmes. Among the tasks involved in the production of numbers, even the classification schemes and their categories are conceived as the product of historically conditioned intellectual constructions. Every statistical objectification procedure implies choices and assumptions that tend to be heavily influenced by the projects and aspirations of their creators—which are not always made explicit. Moreover, all statistics face debates about methods, interpretation, and use. The modes of appropriation and the social uses of the numbers that circulate in the public space are also multiple and diverse, and become understandable when considering them within their contexts. This is the framework in which the chapters of this book can be located, as representatives of a constructivist epistemology that allows opening the 'black box' of statistics to observe what is inside or behind them.

The first section of this introduction examines the interplay between statistics and nation—and state—building processes, looking at how this knowledge managed to remain as a universal science while being developed in national contexts. The following section deals with several of the statistical practices detected in Latin America throughout the period covered by the book allowing us to understand the regional particularities in the assimilation of this science. A third section, relatively connected with the previous one, analyses how the statistical culture dealt with the constitutive social and political frictions of Latin America, shedding light on its role in the hybridisation processes that characterised the history of the region. The last section of this introduction describes the content of the book.

Statistics and the Construction of Latin American Nation States

One of the moments most explored by the socio-political history of statistics—the organisation of statistical offices in Europe during the first half of the nineteenth century—constitutes a period of great political upheaval in Latin America, even of violent confrontation, due to the wars for independence, particularly in Spanish Latin America. These disputes largely stopped the quantitative ambitions of members of the intellectual and political elites and generated a deadlock in the quantifying processes. Political stability seemed a requirement to enumerate. In a certain sense, however, these conflicts also nurtured the interest in counting. For the not yet fully consolidated states, the disputes brought to the forefront the need to strengthen themselves militarily, and hence to know the number of individuals suitable for military service and to ensure the domain of the territory.

Interestingly, these objectives were not very different from those the Spanish monarchy had pursued during the colonial period and particularly during the last decades of imperial rule. Indeed, during the eighteenth century, the metropolis systematically sought to collect news of its overseas colonies. These experiences were valuable precedents for future inquiries. To satisfy this need, the metropolis encouraged the organisation of scientific expeditions and the distribution of questionnaires for the preparation of topographical descriptions. While the expeditions involved organising and financing itinerant commissions of scientist and polymaths to gather and produce knowledge, the groundwork behind the topographical descriptions implied entrusting these investigations to local administrative bodies, including civil, military, and ecclesiastical officials. In these cases, imperial bureaucrats officiated as implementers and informants. Most of those questionnaires were prepared by the Council of the Indies (*Consejo de Indias*) with the aims of knowing the territory, identifying its resources, and determining the conditions and characteristics of the subjects of the monarchy. Colonial authorities were responsible for coordinating the distribution of surveys, processing the information collected, and writing reports.

This practice was not restricted to the Enlightenment years. These inquiries, which in some cases had more than four hundred questions, date from the first days of the Conquest. Over thirty questionnaires crossed the Atlantic during the three centuries of colonial rule—eight of which took place during the eighteenth century. Although some inquiries were designed for specific territories, most of them had a continental focus, imposing relatively uniform processes of data gathering. Also known as land descriptions (*relaciones de tierra*) or geographical descriptions (*relaciones geográficas*), these colonial texts required extraordinary administrative efforts, which mobilised and connected almost all the bureaucratic strata of the Spanish Empire (De Solano 1988).

In that sense, the Spanish monarchy was able to put into motion a massive transatlantic chain of documentary production. To do so, it bundled together a myriad of actors that will also be entrusted to produce statistics in the post-colonial period: the military, parish priests, and influential inhabitants (*vecinos*). Hence, the colonial production of imperial knowledge left significant footprints at different levels, but in particular among intermediate and low-level colonial officials, who acted as compilers, reviewers, and even ghost writers of the local reports sent to viceroyal and peninsular authorities. The long-term impact of these colonial efforts of data collection must be considered when examining the environment in which national statistics started to develop (Estefane 2017).

This connection, however, did not necessarily imply the existence of a perfect continuity between the colonial and the post-colonial ways of knowing. From a formal point of view, there were more differences than similarities between both models. Colonial research responded to occasional initiatives and failed in consolidating routine practices. It was not undertaken by specialised bureaucracies, nor did it differ from the daily tasks of government. The knowledge produced did not have a public orientation, nor was it perceived as an input for open political discussions. On the contrary, the data obtained was usually kept under reserve or total secrecy (Ventresca 1995, 50–54).

After independence, it became increasingly difficult to keep social information under reserve. Elections were one of the political novelties that turned the enumeration of inhabitants into a vital practice, subject to permanent scrutiny and disputes. Indeed, given the demands imposed by the principle of popular sovereignty, it was imperative to know the exact number of inhabitants with the right to vote and to determine the bases of political representation. In more abstract terms, imagining those years of political upheaval allows us to understand the vital need those elites had to generate some order within the chaos. From this point of view, statistics could be thought of as a type of rationality that was sought to be imposed on a troubled territory that was experimenting change. Many of the censuses taken in the first half of the nineteenth century were the result of provincial or city-specific efforts rather than of consolidated central states.[1] They were attempts to order a changing reality and give legibility—in the sense proposed by James Scott (1998)—to disparate and fragmented population settlements within a territory where community and local ties persisted, and where it took time to establish common measurement standards.[2]

The literature converges when highlighting the close relationship on both sides of the Atlantic between the consolidation of the modern state and the official production of statistics. Statistical knowledge was developed concomitantly with the process of shaping the national state, which forces us to conceive of statistics not only as a scientific discipline, but as a branch of government. The historical reconstruction of the Latin American statistical trajectory explored throughout this edited volume contributes to the understanding of the role of statistics in the construction of the modern state. How did statistics become part of the administrative repertoire of Latin American states? As in Europe, where no uniform model of organisation of statistical activity emerged despite the intentions of the promoters of statistical internationalism and the recommendations emanating from their institutions (Randeraad 2010), the evolution of statistics as an administrative practice also differed between Latin American countries. Within the region, it is as difficult to make generalisations as it is to find a single institutional model. The strategies of state and nation construction that the Latin American countries applied throughout the nineteenth century, as well as the uniqueness of the political projects and the relations of force that ended up influencing their trajectory, were diverse. Such divergence clearly determined the forms of institutional organisation, the statistical practices, the objectives of the knowledge production processes, and the themes to be explored using numbers.

Even in the countries that achieved internal pacification of the previously disputed territory, the development of statistics as an administrative

[1] Examples include the incomplete 1813 Chilean census (Silva Castro 1953), the 1835 census of the north-eastern Brazilian state of Bahia (Barickman 2003), and the 1836 Lima census (Peloso and Ragas 2001). Arrioja Díaz Viruell's volume (2016) examines statistics and enumerations implemented at a local level in Mexico during the first half of the nineteenth century.

[2] In some countries, like Mexico, this process extended until the twentieth century, see Vera (2017).

tool was subsequently and constantly involved in political and administrative transformations. In relation to the external recognition of these states, the borders that demarcated the territory that they legitimately intended to dominate and that at the same time contained the population to be enumerated were not yet fully established, as shown by the armed conflicts caused, among other issues, by territorial disputes, such as the War of the Triple Alliance and the War of the Pacific.[3] On the domestic front, the tension between civil and ecclesiastical authorities was characteristic of the nascent republics. The young Latin American states tried, with greater or lesser success, to displace the Catholic Church from the registration and control of the population data that it regularly collected through parishes. In certain cases, as the history of Colombia shows, the states would not achieve this until well into the twentieth century.

During the second half of the nineteenth century and until the First World War, the insertion of the Latin American countries into the international market as producers of commodities made them strongly dependent on foreign trade. Within the international stage, these states occupied a peripheral position in relation to the European cultural metropolises. In most Latin American economies, the creation of markets, the establishment of capitalist social relations, and the adjustment of the productive structure to the primary commodities export-led model consolidated oligarchic states. Minority social groups with great economic power not only had the monopoly of the most profitable productive or extractive activities. They also controlled access to the government as well as the mechanisms of political succession and, therefore, the administration of the state. These countries experienced the great weight of the elites in the national political system. Given the government's appetite to integrate both commercially and culturally into the world, the production of statistics was valued as one of the ways to demonstrate that Latin American countries were following the path of the civilised nations of Western Europe (Loveman 2013). Considering themselves as well as being perceived by the rest of the world as 'modern' was one of the most sought out goals of the newly unified states. In the framework of the "globalisation of figures" movement of the late nineteenth century (Brian 1989; Gagnon 2000), carrying out population censuses as well

[3] In the War of the Triple Alliance, Paraguay fought against the coalition formed by the Brazilian Empire, Uruguay, and Argentina between 1864 and 1870. The War of the Pacific confronted Chile with Peru and Bolivia between 1879 and 1884.

as compiling records of the country's foreign trade or its demographic evolution in the form of tables and graphs was a way of presenting themselves as a member of the club of 'civilised' nations.

In those years, however, statistics was far from being a simple administrative practice carried out exclusively—and depending on the case—by statistical offices, ministerial departments, or other government agencies. It is possible to imagine how difficult it was to carry out statistical research in Latin America before the extension of railway networks, with rudimentary telecommunications, in territories lacking infrastructure, without safe transportation, and without the technology that facilitated arithmetic operations. Chile carried out its first formal national census in 1854. Costa Rica followed in 1864, Argentina in 1869, Brazil in 1872, Venezuela in 1873, Peru in 1876, Guatemala in 1880, and Mexico in 1890. These national population censuses were immense logistical operations given the limited resources available at that time and taking into account the widespread illiteracy of the population. These structural conditions influenced the way statistics were adopted throughout the region.

Likewise, while states struggled to achieve political centralisation as well as territorial penetration, the formation of the national statistical apparatus was thwarted by tensions of various kinds. In certain cases, like in Colombia (Estrada Orrego 2017), the most strained relations were between the central and the local powers. However, in general terms, these rigidities or institutional disputes varied depending on whether countries were politically organised based on a strong—like Brazil—or weak—like Argentina—federalism. Within the first group, the tensions tended to manifest themselves between different levels of government, in the form of clashes between the authority at the central or national level and the states or municipalities controlled administratively by local elites. In the second case, visible tensions prevailed between agencies or departments of the different ministries at the level of the executive branch of government. The disputes took the form of jurisdictional disagreements and bids for the cognitive monopoly on certain statistical objects.

Cultivating Statistics from Some Corner of Latin America

If the production of statistics and censuses emerged in Latin America as an imperative of modernity, it also expressed the desire of the ruling elite to have a handful of certainties to face a changing and frequently agitated reality. Statisticians and those interested in statistics shared the assumption that an underlying order existed, and that statistical language would sooner or later allow them to reveal it. However, their aspiration to find that order was amid the political effort to build nation states. This made Latin American statisticians focus both on the expansion of the corpus of statistical knowledge and on imposing statistics as a guide for state administration and political decisions. In the continuous search to establish the legitimacy of official statistical practice by asserting the relevance and usefulness of the figures, symbolic identifications flourished. Statistics became a compass—the tool required to steer and guide the state vessel. In parallel, the creation of statistical offices was perceived as progress concerning the administrative organisation of the state. Statistics were thought of as a function of what statisticians and those interested in statistics imagined to be a modern and rational government.

These common expectations, however, should not hide the ample variety of profiles that statistical practitioners in Latin America had in this period. In a context in which scientific disciplines were in full bloom in the region, in each country statistics joined their fate with that of the members of a wide range of intellectual, professional, and academic communities. Doctors, lawyers, military men, journalists, engineers, geographers, among others, were called upon to establish and spread the statistical language throughout Latin America, but they did so from very diverse institutional positions and academic communities. Moreover, several Europeans—such as Francisco Latzina in Argentina, Francisco de Bèze in Chile, or Anthony John Schlesinger and Sophus Höeg Warming in Colombia—played a relevant role in shaping national statistical apparatuses in the second half of the nineteenth and in the first decades of the twentieth century. In Latin America, domestic and foreign census takers combined the knowledge, abilities, and skills that they had developed themselves throughout their very different formal and informal training and personal journeys. Beyond some widely known sporadic courses on statistics that existed in certain countries or the superficial assimilation of its categories and methods in the courses of political economy taught at the main institutes and universities, statistical teaching was not rooted in the region. These courses can be conceived as an inaugural yet isolated milestone that was not really a consequence of the institutionalisation of statistical training which acquired continuity in the region years later.[4] Some people even carried out highly significant individual and independent research that fostered the development of statistics in their country. However, it is difficult to separate their role from that of the local academic groups or scientific societies that they formed part of.

What was the role of scientific associations in promoting statistical practices in Latin America at that time? How can their gravitation or influence be accounted for? With the existing knowledge, a regional generalisation would lead to hasty conclusions. Although this edited volume assumes that throughout the period analysed in Latin America official statistical services prevailed over any other private quantification initiative, the existing research does not portray a complete institutional and sociological picture of the spheres in which statistical knowledge was built upon in the region during the nineteenth century. The references that specialists make regarding the coexistence of both official—in the sense of administrative or governmental—and private statistics—that Laura Cházaro highlights for the case of Mexico and Alexandre Camargo points out for the case of Brazil—suggest that the analysis of both state offices and agencies implies a partial understanding of the region's statistics.

As in other parts of the world, in Latin America civil society was also the birthplace of groups that were involved in various ways in the deployment of statistical practices and intervened at different times in highly elastic statistical chains. Without being robust enough to generate their own surveys but being strong enough to demand statistical research from the state, interest groups played a role in promoting statistics in some Latin American countries. Although the historiographic progress made in this regard indicates that interest groups did not completely lead the quantification nor did they unilaterally establish the criteria for the statistical measurements, they did have the ability to block, question, or criticise and, thus, condition official statistical inquiries. Hence, to advance in the quantification of markets or of certain domains of the social world, the state needed to align the goodwill of these different interest groups.

The evolution of statistics in Latin America was also fuelled by debate and controversy. In Mexico, for example, since the 1830s and 1840s "statistical enumerations and calculations were constantly debated [because]

[4] Camargo Pereira and Morettin (1991) for Brazil and Mentz and Yohai (1991) for Argentina, for example, analyse the trajectory of statistical education.

different valuations and images of the population were derived from what was measured. To create a single nation, it was necessary to discuss which of all these possibilities was the best. The discussion was long and lasted for the whole of the nineteenth century" (Cházaro 2001, 41, authors' translation). In Brazil at the beginning of the twentieth century, sanitary physicians engaged in important debates with government statisticians (Camargo 2007), while in Chile contemporary observers were able to criticise the census figures on occupation and endorse them with scepticism (Hutchison 2000). In Argentina, resistance to as well as questioning and criticism of official statistics even led to the interrogation of certain representative individuals of the statistical civil service who engaged in public discussions with interest groups such as industrialists and political groups like the anarchists (Daniel and González Bollo 2010).

During the second half of the nineteenth century, a trans-nationalised statistical community was consolidated given the regularity of events such as the International Statistical Congresses that took place in different European cities and with the subsequent creation of the International Statistical Institute (Brian 1989). Despite their irregular assistance to these meetings, Latin American statisticians remained connected with those institutional gatherings of statistical internationalism. Together with other communication channels such as journals, these spaces for interaction and exchange between state and academic statisticians from different countries of the world generated an international network for the circulation of ideas (Gagnon 2000). From those congresses where enlightened decisions were made, there was a proliferation of regulations and guidelines on statistical work, proposals for institutional setups, and classification schemes for administrative or census statistics, which expressed a series of efforts aimed at standardising international statistical activity. Some scholars that study statistical internationalism of the late nineteenth century emphasise the type and content of these norms, as well as the rules and values that sought to govern this trans-nationalised statistical community. Others point out that, above its standardising aim, the organisation of government statistics and national concerns influenced the international agenda (Randeraad 2010, 8). To the extent that they tended to bring the conventional aspect of statistical practices to the forefront, these studies are less inclined to give visibility to the internal power structures of those spaces, to identify mechanisms of authority, or to record social dynamics of influence that were not unique nor necessarily unidirectional. The divergent and often contradictory interests of the individuals of each local statistical community undoubtedly deserve more attention. Perhaps this way it will be less difficult to capture the role of Latin American statisticians within statistical internationalism of the late nineteenth and first half of the twentieth centuries as it is known that they have built various bridges both personally—through the exchange of letters—and institutionally—by sending representatives to those events. Through the study of these exchanges, the historiography will reach a greater understanding of their role.

Although Latin American statisticians made use of foreign statistical models and embraced theories perceived in these fields as universal, these experiences never implied a passive reception, or an uncritical assimilation of statistical conventions sealed outside the continent. Latin American statisticians and statistical practitioners also pursued their own cognitive interests and raised original questions that were discouraged by the existing international standards. Even knowing the suggestions and the 'good practices' envisioned in Europe, sometimes Latin American statistical practitioners justified the infeasibility of applying these recommendations to the reality of their countries, constructed independent criteria, and made autonomous decisions. Distancing ourselves from interpretive models that propose a linear influence on local practices or an uncritical appropriation of foreign ideas or notions, it is important to consider statistical internationalism from another angle. From the point of view of Latin American statisticians, these spaces constituted, at the same time, resources available to be put into play at the local level, stamps of prestige, or sources of authority. Sometimes the references of Latin American statisticians to international norms, congresses, and organisations became a strategic resource to strengthen their position within the local sphere given that political elites venerated Europe, to acquire funding, and to innovate organisationally by borrowing the prestige of those institutions and their best-known individuals. On other occasions, also strategically, these ties with the international community were mobilised by Latin American statisticians and statistical practitioners to find external support for local statistical reforms when they were slow or delayed by the absence of government authorisation.

Statistical Culture in the Frame of Latin American Hybridisations

Blending the different populations given the ethnic and cultural diversity that characterised Latin America to generate a national entity was the common challenge for all the countries in the region. The numerical description of their territories and populations was a way of creating national imaginaries. Statistical language not only made it possible to unify an artificially delimited geographic space. It also projected a certain image of cohesive nations on politically fragmented territories and on ethnically diverse population groups. This image of the national entity as a political and demographic unit was supported by knowledge considered positive, that is, numerical, with effects on reality. Thus, abstract entities such as the nation became concrete thanks to statistical aggregations. It is in this sense that the literature highlights that the modern population censuses had a 'constitutive' function concerning the nation, parallel to the political role that they acquired in the configuration of the new republics. According to Mara Loveman (2014), during the nineteenth century, Latin American censuses were guided by two complementary political projects: a descriptive one, which contributed to defining the cultural boundaries of an imagined community, and a prescriptive one, which established racial miscegenation as a positive singularity of Latin American countries when compared to the rest of the world.

Although they shared the political function of being nation-building tools, the region's censuses were not completely identical, at least until the first regional coordination initiatives emerged in 1950 with the aim of conducting the Census of the Americas.[5] During the nineteenth and the first half of the twentieth centuries, Latin American countries did not include the same questions on their census schedules, nor did they even maintain them over time. The Argentine census grid, for example, disregarded questions about the colour or race of its inhabitants—as Hernán Otero analyses in this volume—about the origin of their ancestors, and about religion. These lines of enquiry were, however, pursued by other countries, even by those with similar experiences of receiving massive

[5] In 1940, the Inter-American Statistical Institute was created. Since 1943, its work has been published in the journal *Estadística*. This institute began a series of Inter-American Statistics Congresses in 1947 and 1950 that were later taken over by the Organization of American States and were renamed Inter-American Statistics Conferences.

immigration, such as Brazil. Moreover, the Brazilian censuses carried out in 1872 and 1890 included the racial classification of the population, while the censuses of 1900 and 1920 omitted it. In the Mexican case, the first modern censuses abandoned the racial categories used in colonial enquiries and normalised the statistical categories with reference to 'citizenship' (Cházaro 2018). As Loveman (2009) shows when constructing a comparative matrix, there was substantial variation in the census schedules and in the statistical descriptions of populations contained in Latin American censuses in this period. Therefore, the paths that led to the representation of the population as a homogeneous entity were diverse. In some cases, language and religion were crucial variables, in others racial classifications were prioritised, while in several places nationality was more decisive.

As the bibliographic essay that closes this edited volume examines, censuses and census categories, particularly racial and ethnic ones, were headline topics within Latin American historiography.[6] Exploring the census classifications in historical perspective sheds light onto the constructed character of the populations. The categories defined by census takers, developed based on certain implicit assumptions and racial beliefs, gave meaning and materiality to the notion of population. At the same time, as Benedict Anderson (1983) suggests, these categories became identifiable for the population that they only claimed to "describe".

During the nineteenth and early twentieth centuries, Latin American statisticians adopted classifications that allowed them to imagine a national community based on fragmented populations, while discarding others. In some countries, the census deliberately promoted the ethnicisation of the population's perception by favouring or giving centrality to racial classification (Camargo 2009). There were many differences in the production of racial statistics across Latin America. However, Loveman identifies points of convergence in the conception of 'whiteness' that

[6] This Latin American academic output must be framed considering the interest generated by censuses and racial classifications within the socio-political history of statistics carried out in Europe, the United States, and Canada. The study of classifications—from their definition to their application—followed two trajectories. One was closely linked to intellectual history, focusing on conceptualisations, aspirations, and projects of the men behind the censuses. The other, centred around the history of the empirical classification practices during the enquiry, examined the characteristics of the different actors, resources, and instruments involved in the census.

informed census-taking in the region (Loveman 2009, 210). Undoubtedly, as Cházaro (2001) suggests, the way in which the question of race was intertwined with the numbers provided by official statistics constitutes a Latin American singularity. Behind the concern to determine the 'true' total number of inhabitants, a series of moral notions was at stake about the various elements that made up that problematic unit called population. Throughout their history, Latin American countries have been configured as hybrid cultures, the result of intercultural crossings and mixtures of heterogeneous social groups. Those contemporary statistics contained pre-established judgements regarding these diverse groups that they later projected as self-evident truths. Numbers became the carrier that fostered a moralisation directed both to the homogenisation of the population and to its 'civilisational normalisation'.

In addition to intercultural mixtures, the hybridisation of Latin American societies encompassed the intertwining of the traditional with the modern. As Néstor García Canclini points out, in Latin America there were processes of national construction or state building in which modernising projects have coexisted "with traditions that are not very compatible with what Europeans then considered as characteristics of modernity". The rationalisation of social life coexisted in Latin America "with religious and ethnic fundamentalisms, illiteracy, and archaic power arrangements" (García Canclini 1997, 111, authors' translation). Hybridisation also reached statistical practices. From the institutional point of view, it related to the mode of organisation of statistical practice where there was often a forced coexistence between state actors and ecclesiastical authorities, or the reliance on parish priests to achieve success when carrying out the census. From the point of view of its content, it correlated with the signs of survival of colonial statistics, although many times re-signified, as well as with certain local traditions in the consolidation stage of modern states. Following Camargo (2018), the convergence of modernity and traditionalism in end-of-the-century census practice emerges as a unique feature of Latin America. Properly modern aims that pursued both exhaustive population counts and the compilation of statistics on administrative sources—conceived then as indicators of the progress made by the new nations—were supported by conservative judgements and conceptions. If the standardisation of social life is a properly modern trait, it would be impossible to deny that in Latin America statistics promoted this process. But the hybrid character of their societies was also reflected or translated in their statistics.

Beyond censuses, the statistical production of Latin American states was based on a wide range of practices that aimed to count, classify, and qualify people and things. Why did the Latin American states collect statistics? For what purposes? What were the topics to be accounted for? What political concerns did their statistical programme reflect? Certain historical events, such as the epidemics that hit large cities, were experiences that influenced the trajectory of national statistics by enhancing the relevance of mortality data and numbers on the "natural" characteristics of the population. At the same time, they gave way to the prominence of physicians and hygienists within the national statistical community of Brazil, Mexico, Argentina, and Colombia. An unintended consequence of capitalist modernisation and growing urbanisation—the emergence of the "social question"—attracted the focus of national statistics in certain Latin American countries. At the turn of the century, socio-labour statistics were developed in the light of these new concerns that became a crucial worry for ruling elites. As countries, such as Argentina and Chile, intensified their industrialisation processes, this quantifying impulse gained more force, reconfiguring the map of labour relations and the crucial points of social conflict. Whether it was a health or a labour issue, in both cases the result was the creation of 'social facts' using statistics. By quantifying them, certain perspectives on reality became indisputable facts.

During the first decades of the twentieth century, some Latin American states started producing statistical knowledge on the living conditions of the working class. The quantification of work-related accidents, strikes, occupational diseases, and shedding light onto female and child labour became part of a diversifying statistical agenda. In this scenario, there was a renewal of the toolbox available to official statistics with the development of new surveys, indices, and indicators. Many of these statistical series emerged with the aim to promote the technical management of social tensions—many entrenched around labour relations—in societies that were experiencing profound changes and contradictions. There was the underlying idea that statistics could be a basis for solving urgent social problems. In some Latin American countries, statistics accompanied the state's progress in regulating labour relations as a tool for testing and constructing scientific facts that justified the implementation of laws and social welfare protection. Hence, the deeply rooted liberal political consensus began to weaken by the force of statistical reasoning. The quantification of the labour world contributed to certain situations being understood as social risks, shared by specific groups, and linked to the worker's condition, breaking with an entire individualistic tradition behind the conception of society. The insurance approach and the introduction of social security against risk were somewhat intertwined with statistical language. Socio-labour statistics helped to shape social policy understood in terms of public management of social risks that brought about the consolidation and expansion of capitalist production relations in peripheral countries such as those in Latin America, characterised by deep social and economic inequalities. Although in some Latin American countries statistics traced the evolution of the state's social intervention, much remains to be learned regarding their role in the generation and development of the social welfare models that appeared in the region in the second half of the twentieth century. Given the unique characteristics assumed by the Latin American welfare states, one would expect to find in them aspects, dimensions, and tensions not tangible in other settings.

About This Book

This book brings together works of well-known scholars who research the history of Latin American statistics by combining the study of statistical institutionalisation processes in multiple national scenarios with the global development of the discipline between the late eighteenth and the first half of the twentieth centuries. By compiling a relatively dispersed academic production, the volume makes available to new audiences a variety of interpretations of knowledge expansion processes and statistical practices in Latin American nations. Each of these individual works can be thought of as a piece of a broader socio-political history of Latin American statistics. Despite finding common ground in the colonial past or showing affinities in terms of the economic-structural conditions, these histories portray a rich diversity in terms of political-institutional cultures and state configuration processes, as well as variety in the trajectories of the national statistical apparatuses. Likewise, they explore the (ir)regularity of the census inventories, the statistical instruments that were developed, the political and cognitive interests that modified it, and even the grid of classification of the population proposed in the censuses. Although the link between state and statistics is unavoidable and undeniable, this volume does not seek to offer a history of the construction of Latin American states through numbers. On the contrary, it aims to show the multiplicity of records that traced the practices of data collection and instrumentation of enumerations, and of elaboration and dissemination of statistics within Latin American countries.

The book begins with the chapter "The 'Philosophical Eye' and the 'Industrial Spy'. Statistical Thinking in South America After Independence" by Marcelo Somarriva and Andrés Estefane. Contrasting a series of testimonies by intellectuals, politicians, diplomats, and reformists living in Peru, Argentina, and Chile in the decades immediately after independence, this contribution explores the meanings of statistics in an environment of rapid institutional change. According to the authors, the allegories of the 'philosophical eye' and the 'industrial spy' represent two different approaches to statistical knowledge in this period. The 'philosophical eye' was linked to the reformist projects sustained by the Creoles, by means of which they processed their moral critiques to the worst legacies of the imperial administration. The 'industrial spy' related to an extractive understanding of the territory and its resources, in tune with the mindset of diplomats and foreign investors eager to accumulate critical data for guiding the exploitation of these now independent countries. This distinction, Somarriva and Estefane claim, is crucial to understand the creeping abandonment of the moral vision of statistics and its replacement by a more pragmatic approach. It not only determined the profile of the future statistical offices that will emerge in the following decades, but also the underlying conditions of a sort of new colonial pact with the global powers.

In the following chapter, entitled "Mapping Numbers: Statistics, Cartography and the Making of National Space in Brazil", Alexandre Camargo examines the period that extends between the emergence of statistical cartography in Brazil, including its break with the cartographic tradition of Imperial Brazil (1822–1889), and the reconfiguration of the national territory in the first decades of the twentieth century, as part of its regionalisation. Beyond the fact that, as Nico Randeraad (2010, 2) underlines, during the nineteenth century geography and statistics were disciplines that easily overflowed into one another, this chapter shows how maps and statistics converge politically for the imagination and creation of a common space. Although there are authors who explored these crossroads for other territories (Palsky 1991; Labbé 2010) transcending the relatively independent channels through which the historiography of statistics and of cartography have been developed, Camargo's work is a novel contribution when concerning Latin America. Undoubtedly, with this original perspective, this chapter enriches the historiography of Latin

American statistics by outlining an explanation of the existing regional division of Brazil.

In the chapter "Portraits for an Exhibition: The Making of a Statistical Culture for Public Life in Mexico During the Time of the *Dirección General de Estadística*, 1882–1922", Laura Cházaro analyses how statistics gradually became part of the administrative life of the Mexican state in the period 1882–1922. This process full of obstacles was not linear nor cumulative. Cházaro highlights not only the disagreements between civil servants and citizens, but also the resistance of state officials themselves to carry out tasks such as collecting statistics that were still lacking legitimacy. In this way, the chapter retraces the 'naturalisation' of the association between the state and statistics to show it as emerging from historical contingency.

Within the framework of this edited volume, Cházaro's methodological approach has the particularity of combining elements of institutional and cultural history with a biographical perspective. Antonio Peñafiel personifies a (statistical) gentleman. Cházaro shows how Peñafiel's abilities—not necessarily related to arithmetic or mathematics but rather a consequence of his activities as a collector and an antique dealer—were at the root of the production of official statistics in Mexico at the turn of the nineteenth century. Far from technical skills or competencies, during this period the production of Mexican official statistics relied on personal negotiations and political ties that mimicked the dynamic of the relationships between different elites. Thus, while promoting statistical standardisation or technical mechanisation, these relationships reproduced said system of affections and hierarchies.

In the next chapter, "Ethnic Origins, Race, and Nation in the Argentine Censuses, 1869–1914", Hernán Otero analyses the symbolic role of the first three national population censuses taken in 1869, 1895, and 1914 in the construction of a specific image of the Argentine nation. Following the path outlined by Benedict Anderson and Bernard Cohn on the link between statistics and the nation in general—and with respect to the censuses as instruments for the construction of national populations in particular—Otero examines the symbolic construction implicit in the Argentine censuses, scrutinising the different operations carried out by census takers. First, he studies the criteria established for the construction of the data and the definition of the categories adopted for the classification of the population. Then, he analyses the interpretive reading of the census results. Finally, he explores how data was presented in tables or figures, conceiving them as another discursive dimension. From the specific and detailed analysis of the enumeration of the indigenous, black, and European/foreign populations, Otero's chapter allows us to understand how the Argentine population was conceived of during the formation phase of the national statistical apparatus through the lens of what he conceptualises as the national census paradigm. However, distancing himself from positions identified with extreme relativism, Otero does not attempt to show the 'invention' of the nation through numbers as a fickle or merely arbitrary product, nor does he claim that censuses are photographs or direct reflections of reality. Both the choices, assumptions, and dispositions of census takers and their conceptions embedded in the cultural, political, and ideological framework of turn-of-the-century Argentina implied that the image of the nation generated by official statistics ends up being somewhat partially autonomous from the historical process itself and, at the same time, unique with respect to the 'invention' of the nation *vis-à-vis* other Latin American countries.

Victoria Estrada Orrego's chapter "Loose Numbers: Political Centralisation and Statistical Fragmentation in Colombia, 1886–1930" re-examines two highly significant issues. The chapter uses the official statistical practices as a peephole to observe the changing outlook of the administrative reality of Latin American states during the second half of the nineteenth century. Moreover, using Colombia as a case study, the chapter examines the coexistence between traditionalism and modernity.

This chapter portrays, on the one hand, the trajectory of a state that is in line with its regional newly unified peers in the search to modernise itself by incorporating modern tools as statistics and showing its products, such as censuses and statistical yearbooks, as symbols of modernity. On the other hand, Estrada Orrego's research depicts the path of a state that does not have a secular registry of its population, and thus is far from completing the process of secularisation. During the conservative republic (1886–1930), the Colombian state failed to monopolise birth, marriage, and death records of its population. Hence, it was forced to establish an organisation to produce vital statistics based on parochial certificates and church authorities. The long survival of the traditional parish registries had a significant effect, for example, in terms of mortality statistics and the classification of its causes. It took many years to establish the authority of medical certificates as the yardstick to be followed by local statistical practices. In this sense, Estrada Orrego's chapter allows not only to vindicate the close relationship between the evolution of official statistics and the institutionalisation of other disciplines or professional fields, such as that of medicine and physicians. The chapter also leads us inside a particular step of the statistical chain that relates to the classification of causes of death. At this point, the competition between professional physicians and healers emerged, but it was also contested by priests who were necessary participants within the statistical chain.

Like Cházaro's chapter on Mexico, Estrada Orrego also reconstructs the challenges faced by the Colombian statistical apparatus to consolidate itself. However, she places greater emphasis on its successive reorganisations, the recurrence of changes in the denomination of statistical agencies, and their movement between different jurisdictions or ministries. The chapter focuses on the repeated attempts at administrative reform aimed at guaranteeing statistical centralisation that lasted until 1923 within the framework of the Kemmerer Mission. These attempts must be understood in relation to the international debate about the appropriate mode of organisation of public statistics that begun at the end of the nineteenth century in the Global North and that prevailed towards the turn of the century concerning the options of centralisation, decentralisation, or coordination (Prévost and Beaud 2012, 75–82). The analysis of the Colombian experience not only refines the assumption that there is a single appropriate rationale for the organisation of statistical tasks. This case study also contextualises the principle of statistical centralisation while raising awareness of the variety of meanings and local adaptations that such model historically acquired.

The volume ends with the chapter "Socio-Political History of Latin American Statistics: A Bibliographical Essay", which assesses the existing works on the socio-political history of Latin American statistics and dialogues closely with this introduction. We identify the different paths that this approach has followed and the depth of the scholarly contributions. In doing so, we highlight future areas of research. The essay is followed by a bibliographical inventory of all the works known to the editors of this book that focus on the history of Latin American statistics, several of which are commented upon in the essay.

* * *

In recent years, the progress made within the field of the socio-political history of Latin American statistics has attracted the interest of a group of scholars from different disciplines, extending from Mexico to Chile and Argentina. This community has been increasingly converging thanks to common concerns and it has been working very hard day after day to strengthen its ties. This book is the outcome of that network that has been weaving and expanding, as well as of the solidarity that characterises it. The editors of this volume deeply appreciate the interest and commitment of the contributors of this book, and the rigour with which they have faced this challenge, particularly during COVID-19 times. We also like to thank the *Boletín del Instituto de Historia Argentina y Americana Dr. Emilio Ravignani* (Universidad de Buenos Aires, CONICET) for allowing us to reproduce an updated version of one of its articles. As editors, we are also very grateful to Maxine Molyneux for the trust she has placed on us and in this project. Two anonymous referees have made valuable suggestions to this edited volume. The editorial team at Palgrave Macmillan has been extremely patient at a very difficult time, and the editors are thankful for their continuous understanding. This book could not have been possible without the support of the academic and scientific institutions we belong to and/or fund our research: the Consejo Nacional de Investigaciones Científicas y Técnicas (CONICET, Argentina), the University of Warwick (United Kingdom), and the Agencia Nacional de Investigación y Desarrollo (ANID, Chile).

References

Anderson, Benedict, 1983. *Imagined Communities. Reflections on the Origin and Spread of Nationalism*. London: Verso.

Arrioja Díaz Viruell, Luis Alberto, ed. 2016. *Registrar e imaginar la nación: la estadística durante la primera mitad del siglo XIX*. Zamora: El Colegio de Michoacán.

Barickman, Bert J. 2003. "Reading the 1835 Parish Censuses from Bahia: Citizenship, Kinship, Slavery and Household in Early Nineteenth-Century Brazil." *The Americas* 59 (3): 287–323. https://doi.org/10.1353/tam.2003.0002.

Brian, Eric. 1989. "Statistique administrative et internationalisme statistique pendant la seconde moitié du XIXe siècle." *Histoire & Mesure* 4 (3–4): 201–224. https://www.persee.fr/doc/hism_0982-1783_1989_num_4_3_1357. Accessed 11 February 2022.

Camargo, Alexandre de Paiva Rio. 2007. "Números para o progresso: um panorama da actividade estatística na Primeira República." In *Bulhões Carvalho, Um Médico Cuidando da Estatística Brasileira*, edited by IBGE,

Centro de Documentação e Disseminação de Informações. Rio de Janeiro: IBGE. https://biblioteca.ibge.gov.br/visualizacao/monografias/GEBIS%20-%20RJ/ColecaoMemoriaInstitucional/11-Um%20Medico%20Cuidando%20da%20Estatistica-Bulhoes%20Carvalho.pdf. Accessed 28 September 2020.

———. 2009. "Mensuração racial e campo estatístico nos censos brasileiros (1872-1940): uma abordagem convergente." *Boletim do Museu Paraense Emílio Goeldi* 4: 361–385. https://doi.org/10.1590/S1981-81222009000300002.

———. 2018. "O censo de 1872 e a utopia estatística do Brasil Imperial." *Revista de História da Unisinos* 22 (3): 414–428. http://revistas.unisinos.br/index.php/historia/article/view/15474. Accessed 8 January 2021.

Camargo Pereira, José Severo de, and Pedro A. Morettin. 1991. "Las estadísticas brasileñas y la enseñanza de la estadística en Brasil." *Estadística Española* 33 (128): 559–574. https://www.ine.es/ss/Satellite?blobcol=urldata&blobheader=application%2Fpdf&blobheadername1=Content-Disposition&blobheadervalue1=attachment%3B+filename%3D128_8.pdf&blobkey=urldata&blobtable=MungoBlobs&blobwhere=564%2F582%2F128_8%2C0.pdf&ssbinary=true. Accessed 28 September 2020.

Cházaro, Laura. 2001."Imágenes de la población mexicana: descripciones, frecuencias y cálculos estadísticos." *Relaciones. Estudios de Historia y Sociedad* 22 (88): 15–48. https://www.redalyc.org/articulo.oa?id=13708802. Accessed 28 September 2020.

———. 2018. "Médecins, statistique et recensement: les raisons d'une entente cordiale." *Histoire et Mesure* 33 (2): 33–60. https://doi.org/10.4000/histoiremesure.7970.

De Solano, Francisco, ed. 1988. *Cuestionarios para la formación de relaciones geográficas de Indias, siglos XVI/XIX*. Madrid: Consejo Superior de Investigaciones Científicas, Centro de Estudios Históricos, Departamento de Historia de América.

Daniel, Claudia and Hernán González Bollo. 2010. "Las estadísticas oficiales en la prensa escrita porteña (Argentina, 1890–1930)." In *Estatísticas nas Américas. Por uma agenda de estudos históricos comparados*, edited by Nelson de Castro Senra and Alexandre de Paiva Rio Camargo. Rio de Janeiro: IBGE. https://biblioteca.ibge.gov.br/visualizacao/livros/liv44323.pdf. Accessed 28 September 2020.

Desrosières, Alain. 1998. *The Politics of Large Numbers. A History of Statistical Reasoning*. Cambridge, MA: Harvard University Press.

Estefane, Andrés. 2017. "Elusive Numbers. State Knowledge and Bureaucratic Organization in Chile (1750–1911)." PhD dissertation, State University of New York at Stony Brook. http://hdl.handle.net/11401/78299. Accessed 28 September 2020.

Estrada Orrego, Victoria. 2017. "¿Cuántos somos? Una historia de los censos civiles y de la organización de la estadística en Colombia en la primera mitad del siglo XX." *Historia Crítica* 64: 141–160. https://doi.org/10.7440/histcrit64.2017.08.

Gagnon, M.-A. 2000. "Les réseaux de l'internationalisme statistique (1885–1914)." In *L'ere du chiffre. Systémes stastiques et traditions nationales*, edited by Jean-Pierre Beaud and Jean-Guy Prévost. Québec: Presses de l'Université de Québec.

García Canclini, Néstor. 1997. "Culturas híbridas y estrategias comunicacionales." *Estudios sobre las Culturas Contemporáneas* 3 (5): 109–128.

Hutchison, Elizabeth Quay. 2000. "La historia detrás de las cifras: La evolución del censo chileno y la representación del trabajo femenino, 1895–1930." *Historia* 33: 417–434. https://doi.org/10.4067/S0717-71942000003300009.

Labbé, Morgane. 2010. "Les frontières de la nation allemande dans l´espace de la carte, du tableau statistique et de la narration." In *Les espaces de l´Allemagne au XIXe siècle: frontières, centres et question nationale*, edited by Catherine Maurer. Strasbourg: Presses Universitaires de Strasbourg.

Loveman, Mara. 2009. "The Race to Progress: Census Taking and Nation Making in Brazil (1870–1920)." *Hispanic American Historical Review* 89 (3): 435–470. https://doi.org/10.1215/00182168-2009-002.

———. 2013. "Census Taking and Nation Making in Nineteenth-Century Latin America." In *State and Nation Making in Latin America and Spain. Republics of the Possible*, edited by Miguel Ángel Centeno and Agustín Ferraro. New York: Cambridge University Press. https://doi.org/10.1017/CBO9781139342667.

———. 2014. *National Colors. Racial Classification and the State in Latin America*. New York: Oxford University Press. https://doi.org/10.1093/acprof:oso/9780199337354.001.0001.

Mentz, Raúl, and Víctor Yohai. 1991. "Sobre la historia de la enseñanza de la estadística en las universidades argentinas." *Estadística Española* 33 (128): 533–558. https://www.ine.es/ss/Satellite?L=0&c=INERevEstad_C&p=1254735226759&pagename=ProductosYServicios%2FPYSLayout&_charset_=UTF-8&cid=1259925029848&submit=Ir. Accessed 28 September 2020.

Palsky, Gilles. 1991. "La cartographie statistique de la population au XIXè siècle." *Espace, populations, sociétés* 9 (3): 451–458. https://doi.org/10.3406/espos.1991.1487.

Peloso, Vincent, and José Ragas. 2001. "Estadísticas y sociedad en el Perú poscolonial: el desconocido censo de Lima de 1860." *Histórica* 25 (2): 275–293. http://revistas.pucp.edu.pe/index.php/historica/article/view/8720. Accessed 28 September 2020.

Prévost, Jean-Guy, and Jean-Pierre Beaud. 2012. *Statistics, Public Debate and the State (1800–1945). A Social, Political and Intellectual History of Numbers*. London: Pickering & Chatto. https://doi.org/10.4324/9781315655161.

Randeraad, Nico. 2010. *States and Statistics in the Nineteenth Century: Europe by Numbers*. Manchester: Manchester University Press. https://doi.org/10.7765/9781526147530.

Scott, James. 1998. *Seeing Like a State. How Certain Schemes to Improve the Human Condition Have Failed*. Yale: Yale University Press.

Silva Castro, Raul. 1953. "Introducción." In *Censo de 1813: levantado por Don Juan Egaña, de orden de la Junta de Gobierno formada por los Señores Pérez, Infante y Eyzaguirre*, by Archivo Nacional. Imprenta Chile: Santiago. http://www.memoriachilena.gob.cl/602/w3-article-8952.html. Accessed 28 June 2021.

Ventresca, Marc Joseph. 1995. "When States Count: Institutional and Political Dynamics in Modern Census Establishment, 1800–1993." PhD dissertation, Stanford University.

Vera, Héctor. 2017. "Counting Measures: The Decimal Metric System, Metrological Census, and State Formation in Revolutionary Mexico, 1895–1940." *Histoire & Mesure* 32 (1): 121–140. https://doi.org/10.4000/histoiremesure.5780.

CHAPTER 2

The 'Philosophical Eye' and the 'Industrial Spy': Statistical Thinking in South America After Independence

Marcelo Somarriva and Andrés Estefane

Statistics was a key discipline in the processes of state formation in Latin America. As foundational knowledge for governmental action, it gained an indisputable reputation within the institutional choreography of the nascent republics. The idea, however, was not entirely novel. From the *Conquista* onward, the Spanish Crown and the Catholic Church sought to produce ambitious inventories of the material and human resources available in the New World (De Solano 1988). Going further back, the *quipus* of the Inca Empire emerge as the most sophisticated precedent for this type of knowledge (Quilter and Urton 2002; Salomon 2004). None

M. Somarriva
Faculty of Liberal Arts, Universidad Adolfo Ibáñez, Santiago, Chile
e-mail: marcelo.somarriva@uai.cl

A. Estefane (✉)
Independent Researcher, Santiago, Chile

C. T. Lanata-Briones et al. (eds.), *Socio-political Histories of Latin American Statistics*, Studies of the Americas,
https://doi.org/10.1007/978-3-030-87714-9_2

of these experiences would qualify as a direct precursor of modern statistics, but they did represent significant enumeration efforts—of peoples, facts, and objects—carried out by political structures with high levels of institutionalisation.

Only towards the second half of the eighteenth century, under the influence of the Enlightenment and the Bourbon Reforms, this type of research began to take the form that would eventually lead to the contemporary profile of this science. It is true that a few links were still missing for witnessing the emergence of standardised research programmes or the establishment of complex informational systems entrusted to specialised bureaucracies. But it was in those efforts that the key ideas began to take shape. Statistics was an intellectual innovation as well as a constituent element of the emerging modern state (Westergaard 1969; Woolf 1989; Otero 2006).

The growing demand for demographic, economic, and fiscal information by the Spanish Empire imposed the assimilation of administrative protocols and routines in which many Creole bureaucrats were trained. After the Napoleonic wars and the collapse of the imperial order, that same generation led the efforts to produce knowledge under the auspices of the new republics, governed by the same elites they were part of. The need to know in detail the material resources at their disposal and to determine the total population scattered throughout the territories—crucial for the acclimatisation of the political novelty called the 'principle of representation'—was the basis of the first attempts to institutionalise the production of official numbers. Behind that urge there were moral concerns about how that population lived and what had to be done to improve their situation. That was certainly the response to a shared diagnosis among contemporaries as well as a way of processing the harsh critiques of this generation to what they understood were the regrettable legacies of the imperial rule.

This journey, of course, was not without its setbacks. The complex political scenario left by the wars of independence, the lack of material resources, the conflicts over the physiognomy that the new states had to adopt, and the lack of basic scientific knowledge within the emerging bureaucracies were among the first difficulties for the reception of modern statistics. But there also were profound dilemmas. One of them derived from the need to prioritise between the republican aspiration to produce knowledge for improving the nations and to sustain the new representative politics, and the material pressures to direct those efforts to the production of economic reports for foreign investors. In the end, it all seemed reduced to delineate portraits of stable governments and mature populations that did not yet exist. But no country could avoid this dilemma and, despite the material and political shortcomings when constituting specialised offices, the statistical promise ended up being part of the republican institutional horizon.

This chapter explores the various meanings of statistics in South America, specifically in Argentina, Peru, and Chile, in the decades of formation of the first autonomous political institutions. The political and military imbrication of the independence processes of these three countries gave shape to a space of permanent circulation of people and ideas. Not surprisingly, most of the testimonies quoted here came from intellectuals, politicians, and reformists who understood and lived the independence as a regional experience. This chapter looks at that circulation with the aim of analysing what was at stake—in terms of expectations but also of political action—in the first modulations of the statistical language. Statistics has never been a homogeneous scientific field, and at least during the nineteenth century it was difficult to stabilise a single definition of what the discipline was (Randeraad 2010). This is particularly pertinent to the post-imperial context when key definitions were still open to debate.

Following that line, this chapter proposes that the statistical thinking of this period could be better characterised when distinguishing between two basic approaches. On the one hand, a moral understanding of statistics, a vision semantically attached to the projects of political and social regeneration promoted by Creole intellectuals who, since the last decades of the colonial period, endorsed the adoption of this science as symbol of patriotic zeal. On the other hand, there was the pragmatic vision of this knowledge, more related to the description of the land than with its improvement, and whose quantitative leaning would prefigure the bases of an extractive understanding of the territory and its resources. Likewise, both ways of understanding statistics referred to different schemes of institutionalisation and development, the former connected with the field of political economy and the idea of investigation as the individual enterprise of well-trained polymaths, and the latter with properly organised bureaucracies, where processes of knowledge production depended on formally established state agencies.

Both visions are encrypted in the allegories of the 'philosophical eye' and the 'industrial spy', which were the images that contemporaries used to represent diverging approaches towards the lands and its peoples and different conceptions of the value and uses of scientific knowledge. This chapter does not propose that there was an overt conflict between these two assessments, which most of the time seemed to overlap under the pressure for producing up-to-date knowledge or were expressed by the same people. However, there were semantic differences, which can be read as symptoms of the ideological and political transformations that marked the transition from colonies to republics. In the case of statistics, such transformations will mean the creeping abandonment of the moral understanding in favour of the more pragmatic vision.

This chapter is divided into three parts. The first two are focused on the historical approaches mentioned above. The third and final section offers some closing remarks. The first part gathers proposals and reflections by intellectuals, politicians, and reformists living in Peru, Argentina, and Chile in the decades immediately after independence. These testimonies are grouped together as pieces of an irregular puzzle that account for the meanings of statistics for this generation, the one that developed the first definition of what statistics was or could mean in South America during a foundational political transition. It is through these voices that the section explores the links between statistics and political economy, and the links of both with the idea of 'moral principle', which was the vortex that allowed Creoles to anchor their patriotic zeal in the historical-political critique of the aftermath of the imperial rule. The second part somehow replicates the previous exercise but tracing the testimonies that built a more pragmatic view of this science, emphasising its descriptive role and its quantitative dimension. Of particular interest are the scenes in which statistics appear in dialogue with fields other than political economy, or as part of the politics of construction of foreign relations by capturing the portraits with which these countries confronted—by attracting of defining—the world's gaze.

The Moral Principle

The construction of national statistical systems was one of the main aspirations of Latin American Creole reformists during and after independence. This desire expressed the conviction that it was impossible to think of political economy without statistical knowledge. French economist Jean Baptiste Say, an author of great influence in the region, asserted that this connection was equivalent to the one that related history with politics.

Statistics and history were fixed or diachronic, while economics and politics were dynamic or synchronic. According to Say, statistics and political economy were both sciences, but while the latter made it possible to think about the creation, distribution, and consumption of wealth in general terms, the former made it possible to measure the state of the population and the productive forces in a given period of time (Say 1804, 7–9). In the end, the two formed an inextricable combination. Given the scenario left by the collapse of the Spanish Empire, it was urgent to gather data or 'facts' from scratch. There was no other way to rebuild local economies and organise the future by rationally allocating available resources and productive factors.

Intellectually speaking, statistics was a gigantic achievement that combined different fields of knowledge. It also was perceived as the greatest tribute a patriot could make to his homeland. This was the vision of Peruvian physician and naturalist Hipólito Unanue (1755–1833), one of the first Latin American intellectuals to produce statistical knowledge even before political independence was yet in the horizon. Unanue saw in statistics not only a framework to produce a comprehensive account of the resources, the population, and the industrial and commercial activities of a human community, but also an art to draw an accurate portrait of the political body of a nation (*Mercurio Peruano*, January 1, 1791). This last image acquired an undeniable moral tone in postcolonial Latin America. In the case of Peru and its neighbours, this meant a call to rediscover the territory and its human and material features, following the plans the pioneers of the independence drew up in their writings and proclamations and which aimed to a complete regeneration of these political bodies. This was the way to compensate for the omissions and neglections of the past, naturalised by the indifference of the Spanish administrators, who were neither prepared nor interested in appreciating what seemed evident for the Enlightenment reformist: America was a land of astounding riches. As not everyone was prepared to perceive the extent of these riches, the portrait was also a way to reaffirm the Creole's credentials and skills to govern their own countries.

Unanue's vision of a national survey resonated in the statistical project developed in the pages of the newspaper *El Telégrafo Mercantil* of Buenos Aires as early as 1801. Here statistics was also described as the physical and moral portrait of the political body, as the confluence of geographic and topographic news that would allow to represent the size and capabilities of a fraction of the empire. Indeed, statistics should be the main objective of a projected '*sociedad patriótica*' (patriotic society) based in Buenos Aires, whose efforts should be aimed at constructing the best possible diagnosis "of the natural and political state of the Viceroyalty" (*El Telégrafo Mercantil*, September 2, 1801, authors' translation). Not surprisingly, many of the topics covered by this newspaper foreshadowed the concerns that inspired the organisation of the different national statistical system for years to come. Simultaneously, this knowledge was a requirement for the improvement of the country, a tool for resolving internal controversies, and a universal language to present the new republics to the world. Although the uses and role of statistics changed throughout the nineteenth century, these characteristics remained as a kind of fixed background that captured the anxieties and desires this science was able to merge.

Argentine Bernardo Monteagudo (1789–1825), a key and controversial actor in the independence processes of Chile, Peru, and his own country, also reflected on the usefulness of statistics when recalling his first years in Lima right after independence. In his opinion, the lack of knowledge and reliable data made it impossible for the authorities to glimpse the extent of Peru's wealth and, consequently, to take the appropriate measures to improve its situation. He had no doubt that Peru was one of the richest places on earth, but no one would realise that unless authorities took seriously the truths of 'political arithmetic' (Monteagudo 1823, 16). In the same vein, José de Larrea y Loredo (1780–1830), supposedly the first Peruvian Minister of Finance in putting liberal ideas at the service of Peruvian reality, justified the deficiencies of his 1826 annual report by pointing out to the weak development of political economy, which in turn was a consequence of the lack of statistical information on the country's departments, the main subnational unit. In this regard, for Larrea y Loredo Peru was still far from the civilised nations that had already adopted statistics "to ensure its prosperity" (Larrea y Loredo 1875, 6–7, authors' translation). Coincidentally, that same year the newspaper *El Peruano* (The Peruvian) began to publish a series of studies entitled "Bases para la estadística del Perú" (Bases for Peru's Statistics) authored by the influential landowner Andrés de Salazar y Muñatones. More than raw statistical data, the "Bases" collected some hypotheses and estimates about the demographic and economic situation of the country based on the indicators prepared by European economists. The following year, this same newspaper began to publish some demographic calculations sent by the provincial political authorities of Yauyos, Huarochiri, Cusco, and other administrative units (Salinas 2001).

If Hipólito Unuanue was a kind of pioneer for these first statistical inquiries, he was also a reference for the publication of almanacs and travel guides inspired by the tradition of the inspection visits that were so common during the eighteenth century. Even though these publications were highly popular, and that almanacs and statistics resembled each other, the latter was considered a superior form of national portrait, more exhaustive and vaster in terms of its contents. According to José Antonio Valdés, editor of the Buenos Aires newspaper *El Censor*, statistics moved between history and politics, because it was a way of measuring what the past had created as well as a tool to model what would come:

> Statistics: a word assimilated to portray the political situation of each country. It examines and describes the state of the nation, of its agriculture, industry, and commerce; the means available to promote its improvement; its population, the causes of its growth and decline; how lands are distributed and cultivated, the nature and value of their products; the amount of circulating money, and of properties; the predominant diseases, their causes and cures; people's occupations, which ones must be protected and which ones suppressed; the living conditions of the poor; the state of schools and other public institutions; the state of towns and villages, and what measures may foster their *policía* and improvement; the state of people's habits, religion, science, etc. (*El Censor*, April 24, 1817, authors' translation)

For Valdés, as for many of his contemporaries, statistics was not only a description, but also a sort of 'barometer' of civilisation. This connection derived from its capability to measure the nation's population, which in turn referred to the belief that the progress of a country was tied to the number and strength of its inhabitants, in a physical and moral sense. From an economic perspective, population was synonym of workforce, and the number of people was a way of imagining the number of 'arms' of the political body. Thus, civilisation was proportional to the health of such body, which meant having a country immensely populated by strong, hard-working, and educated people.

Once again, the language of physics and morals converged in the terms similar to those proposed by French *idéologue* Destutt de Tracy, who claimed that a declining population was the real cause of a nation's 'stagnation' (Destutt de Tracy, 1824, 8–9). According to historian Emma

Spary, "during the eighteenth century, a new field of social mathematics was developed to investigate changes in the population of nations, in order to determine how healthy—in moral and physical terms—the nation was" (Spary 1996, 185). History showed that both empires and nations could decay, that barbarism could dismantle civilisation. The fragility of progress was at the centre of the works by Edward Gibbon, Jean Jacques Rousseau, and Constantin de Volney, and their anxieties haunted South Americans as well. For instance, these were the concerns of the Argentine priest and politician Juan Ignacio de Gorriti (1766–1842) when writing his *Reflecciones* [sic] *sobre las causas morales de las convulsiones interiores en los nuevos estados americanos* (Reflections on the Moral Causes of the Convulsions of the New American States). The logic of history was perceptible to the 'philosophical eye', stated Gorriti, the observer's best instrument for measuring the state of a country, its urban conditions, and many other developments (Gorriti 1836, 93–95). In this line of reasoning, enlightened Creoles and their intellectual heirs could not help but think that the declining state of their lands was caused by their depopulation.

But Europeans also taught that barbarism and chaos were avoidable, that society was perfectible, and that humanity had the tools to improve its environment or achieve physical and moral regeneration (Spary 1996, 187). The editors of *El Censor* emphasised this argument in 1817, referring to the declining population of Buenos Aires and to the physical and moral degradation of the city. "Every writer of political economy", the article added, "has affirmed the indisputable dogma that a large population is one of the greatest objectives of the fathers of the people". Only a healthy and robust population would set national industries into motion, opening new sources of prosperity, and allowing citizens to make "unexpected and prodigious" progresses (*El Censor*, May 8, 1817, authors' translation).

All these news help to understand in what terms these debates were treated in the emerging public sphere of South America. In 1822, the government of the Province of Buenos Aires launched the *Registro Estadístico* (Statistical Registry), a monthly journal dedicated to the dissemination of reports on the national territory and to overcome the antiquarian understanding of statistics, which saw it as a mere intellectual adornment or as a source of 'curiosities' for writing local stories. Statistics, on the contrary, was a modern political tool, a source of empirical information that would give life to political economy strengthening its scientific status and helping it to fulfil the high duties that society demanded (Bagú 1966, 205). *Registro Estadístico* had only nineteen issues, but its objectives were taken up by *La Abeja Argentina* (The Argentine Bee), which begun to publish 'statistical news' also with the aim of determining the characteristics (*estado*) of the population. For its editors, statistics was intended to be a social mirror to test the progress of the reform plans and to measure the regeneration of society, looking both at its physical factors—agriculture, industry, commerce—and its moral elements—traditions, habits, religion, and education (*La Abeja Argentina*, December 15, 1822).

Chileans also had the opportunity to read newspaper articles that presented statistical knowledge as irreplaceable both for the exercise of power and for moving towards the future. Speaking about bees—the Napoleonic allegory of industriousness and orderliness—*La Abeja Chilena* (The Chilean Bee), edited by the Peruvian-Chilean politician and writer Juan Egaña in Santiago in 1825, dedicated two of its eight issues to the reproduction of statistical summaries concerning the 'greatest powers of Europe and America' as a way of illustrating the criteria with which the power of a modern state had to be measured (*La Abeja Chilena*, August 20 and 23, 1825). French physician Pedro Chapuis did the same in 1827 from the pages of *El Verdadero Liberal* (The True Liberal), also published in Santiago, by inserting a detailed comparative table that recorded the territorial extension, population, productivity, public debt, and the number of army and marine personnel of the main European monarchies. "Every day we hear lectures about the different states of Europe", he wrote at the top of the table, "but without any other base than the arbitrary will of those who speak. We believe that it would be useful to offer more concrete notions in this regard, and for this we have found a statistical and comparative table of the great powers" (*El Verdadero Liberal*, July 24, 1827, authors' translation).

Colombian Juan García del Río (1794–1856), who after being diplomat and a prolific editor will become president of Gran Colombia for a short period (1830–1831), also reflected on the meanings and usefulness of statistics while living in Chile. In the inaugural issue of *El Telégrafo* (The Telegraph), one of the several newspapers he founded and directed, he recommended Chileans to 'cultivate' the 'little known' science of statistics as a way to be prepared to serve the nation. To contribute to this call, García del Río promised that *El Telégrafo* would publish statistical tables on population, territory, resources, commerce, revenues, and other critical issues for the South American nations (*El Telégrafo*, May 19, 1817). During his brief residency in Chile, another foreigner, Bernardo Monteagudo, used the pages of *El Censor de la Revolución* (The Censor of the Revolution) to also encourage the production of statistics (April 30, 1820). By strengthening the connection with political economy, gathering statistical data became a kind of 'modern imperative' in tune with the spirit of the time and subtle but not entirely away from the patriotic stamp.

A few years later, friar and writer Camilo Henríquez (1769–1825) announced in the first issue of *El Mercurio de Chile* (The Mercury of Chile), the newspaper he had founded upon his return from Buenos Aires, that statistics would be one of the main topics of his new publication. Indeed, this was a demand from Bernardo O'Higgins, the first head of the Chilean state, who wanted Henríquez to produce the "great inventory of the country; the chart of its needs; the state of what exist and what should exist; the indicator of what need to be restored, created, destroyed, modified, and organised". As this was a gigantic demand and the required data would be collected at a slow pace, Henríquez was content to start outlining the script that any future statistical report should follow: population, agriculture, industry, commerce, governmental acts and their impact on society, public finance, credit, and mining (*El Mercurio de Chile*, N° 1 and 2, 1822, authors' translation).

A similar view was held by *El Amigo de la Verdad* (The Friend of Truth), a Chilean newspaper with a particular focus on political economy. The editors claimed that without statistics both political economy and legislation—the foundations of any 'social edifice'—were just mere 'hypotheses'. Prosperity was thus reserved for those states whose governments showed a serious commitment to science, albeit it was not clear what the meaning of that commitment was. The extraordinarily exhaustive list of topics to cover as part of this new statistical project revealed how imprecise that vision was. In addition to ordinary matters—population, agriculture, climate, etc.—the editors advised to record phenomena like the direction of winds and the spending habits of Chilean women (*El Amigo de la Verdad*, April 11, 1823). Symptomatically, this miscellaneous understanding coexisted with others, such as that of the military man Francisco Antonio Pinto, a key figure in Chilean liberalism. In 1825, he pointed out to another critical aspect of statistics and of demography in particular: the relationship between the concentration of inhabitants and the quotas of political representation (*La Década Araucana* [The Araucanian Decade], July 22, 1825).

Considering the magnitude of expectations about the power of statistics, it is easier to understand why the first projects to build national statistical systems had difficulties to materialise. As Henríquez had said, these were titanic commissions and the efforts required to achieve their goals were beyond the capabilities of patriots eager to express their national loyalty. It was becoming clear that these tasks could not depend on the zeal of an individual, but on organised and more professional state offices, whose works in turn had to conform to a different type of research programmes. But this transition could not take shape completely ignoring the moral understanding of statistics embraced by the first generation of Creole reformists and bureaucrats. Venezuelan intellectual Andrés Bello (1781–1865) synthesised the depth of this vision in a late reflection apropos of the growth of the Chilean population and the lack of adequate means to measure it. Statistics was not only about counting people. It also had to do with the aspiration to determine the strength of a country, which was related with fostering the 'moral principle', as described by Bello in *El Araucano* (The Araucanian) in the mid-1830's:

> In any country, the increase in population must be equal to that of the means of subsistence, and in accordance with the development of the intellectual and industrial spheres of the society. If the means of subsistence grow less rapidly than the population, as it happens in countries where the moral principle lacks of the necessary force, the consequence is the misery of the lower classes, and the procession of vices, crimes, diseases, plagues, and of all the means that nature uses to restore balance, such as reducing the average length of life, an undeniable indicator of weakness of the moral principle. (*El Araucano*, January 23, 1835, authors' translation)

For most of the Creoles thinkers of the time—intellectuals and bureaucrats who witnessed the collapse of the Spanish Empire and the rise of a new order—the physical and the moral were inextricably connected. Such a link could be measured looking at the length of life, the extent of vices, and the ubiquity of poverty. Diseases, vices, and poverty were in fact a moral triangle that obsessed Creoles, mostly because they were the legacies of the Spanish administration, but also the symptoms of a mindset of backwardness to be overcome. This is part of what fuelled the reform mentality of this generation, and what turned political economy and statistics into timely scientific tools for healing the 'moral principle', that political equation that had been neglected by the empire and that the republic would supposedly redeem.

Useful Descriptions

Statistics was a knowledge intended not only to remedy problems, but also to collect and disseminate information about the new countries. This was crucial for the formation of a new consciousness in the decades right after independence, one aimed to introduce these nations to the world and establish a new colonial pact with other empires. Statistics became thus an artefact for building foreign relations. Behind this drive there were material, more prosaic needs: to obtain diplomatic recognition and attract immigrants and investors. A book that embodied these aims was *Noticias históricas, políticas y estadísticas de las Provincias Unidas del Río de la Plata* (Historical, Political, and Statistical News of the United Provinces of Río de la Plata) by the Argentine historian and politician Ignacio Núñez (1792–1846) which was written after the suggestion of British Consul Woodbine Parish, who advised that such work would facilitate the signing of foreign trade treaty agreements between Great Britain and the United Provinces of the Rio de la Plata. The ultimate purpose of this work, which was later translated into English and French, was to gather reports on Buenos Aires and the interior provinces in order to display the most seductive portrait of the country to attract foreigners and capital (Núñez 1825). Along with history, statistics began to be used permanently with clear propaganda purposes, and which made it a science with more mundane goals.

Interestingly, at some point it became difficult to distinguish between statistical reports and mirror publications such as travel guides, almanacs, and repertoires (*repertorios*), which were frequently edited in Lima, Santiago, and Buenos Aires. These books were also prepared to please the foreign gaze, but they had lower scientific aspirations and were perhaps more plain in their purposes. But still, it is worth thinking of the fuzzy boundaries between genres that used a similar language and formulated political arguments based on numerical information. A good example is the *Repertorio Chileno* published in 1835 by the Chilean diplomat and politician Fernando Urízar Garfias (1804–1876), who would become director of the first Chilean Statistical Office (*Oficina de Estadística*) in 1843. The main objective of his book was to instruct readers about the situation of the country. The book's size, index, and structure were modelled on any standard almanac, with chapters focused on historical news and the description of the system of government. However, this repertoire was better in its approach to trade and productive activities. This was so because the publication had a clear goal: in addition to providing information, it sought to attract foreigners, who would be 'useful' to the homeland given their knowledge, business, and financial power. That is why Urízar Garfias presented Chile as "a virgin country, favoured by a benign climate, rich in natural resources for working men, inhabited by people of a soft and hospital character (...), governed by liberal institutions that [offered] as many guarantees as [it could be] desired regarding the safety of people and their property" (Urízar Garfias 1835, ii–iii, authors' translation).

Although we have not detected the existence of translations, the promotional tone of Urízar Garfias's text is confirmed by the content of some critical sections: customs and warehousing system, with emphasis on port and import duties; status of foreign residents, specifying the requirements for nationalisation; system of currencies, weights, and measures, with their equivalents abroad; and the list of Chile's diplomatic agents abroad and of foreign representatives in Chile. But at the same time, the text also resembled a modern statistical yearbook, collecting and systematising scattered data. In fact, the essays on general administration and the statistical descriptions of the provinces, the most intensely analysed subjects in the book, certainly managed to provide readers with a satisfactory overview of the progress of the country. Curiously, Urízar Garfias (1835, i–ii) thought of his work as an exercise of statistical description, but a well-known contemporary of him, French naturalist Claude Gay (1800–1873) defined it as a '*guía de forasteros*' (visitors guide) in a letter in which coincidentally recommended Urízar Garfias for leading what would be the country's first statistical office (Gay 1842, 40).

Scientific expeditions, a practice intensely promoted by the new governments, also brought statistics closer to fields other than political economy. Geography was one of them. Statistics, maps, and travel books were thus related results of the scientific commissions that toured South America after independence. These expeditions were seen as a way to rediscover the national territory, but they also aimed to identify the natural resources that could arouse interest abroad. Of course, this practice had a robust history in the continent. Eighteenth century topographical descriptions, the memories submitted to the *Tribunal del*

Consulado (Guild Court) in Buenos Aires, Lima, or Santiago, and even the geographical reports published from time to time in *Mercurio Peruano* (Lima) and *El Telégrafo Mercantil* (Buenos Aires), were some of its precedents. What is interesting to note is the extent to which the intellectuals of the region understood these visits as an opportunity to merge different fields of knowledge—botany, geology, astronomy, political economy, and statistics—into one single vision of the territory.

The first draft of a constitution in Chile offers valuable insights into how statistics was understood as a by-product of the regular inspection of the national territories. Written by Juan Egaña, previously mentioned as editor of *La Abeja Chilena*, this constitution included the creation of a *Consejo de Economía Pública* (Council of Public Economy). In addition to being in charge of promoting the "industrial, rural, and commercial" development of the nation, this institution had to carry out "political arithmetic investigations" and apply its results for "*policía* (policing) and governmental purposes". Two out of the six directors of the *Consejo* had to tour the national territory verifying inspection visits; they were in charge of collecting news about the economy and proposing measures for its improvement. Two others would travel abroad to promote trade agreements and identify reform plans to be replicated in the country. The remaining two, residing in Santiago, would assume the daily administrative routine of the institution (Egaña 1813). Even though this draft was never implemented, some of its chapters were included in the Chilean Constitution of 1823, also written by Egaña. In this second text, the *Consejo de Economía Pública* was renamed as *Dirección de Economía Nacional* (Bureau of National Economy) and 'political arithmetic' was replaced by the modern term 'statistics' (Chile 1823).

In 1821, the editor of the newspaper *El Curioso* (The Curious), invited its readers in Buenos Aires to contribute to the journal by sending scientific news. The purpose of the invitation was to solve practical problems, ranging from the elimination of ants to finding the easiest, cheapest, and safest method to exploit the great Pampas near Buenos Aires. This last image, the economic colonisation of the Pampas, piqued the interest of a fictional reader who sent a letter sharing a fantasy of scientific frenzy, where the Pampas seemed covered by sextants, compasses, and other 'philosophical instruments' used by a legion of enthusiastic surveyors. This imaginative reader also demanded from the government the organisation of an 'expedition of savant' (*expedición de sabios*), like the one

Jefferson had organised in the United States, to draw a geographical portrait of the hinterland and to collect valuable knowledge for commerce, sciences, and the arts (*El Curioso*, N° 2 and 3, 1821). Of course, *The Curious*' fictional reader was not fantasising, but anticipating the immediate future. A few years later, the *Comisión Topográfica* (Topographic Commission) was created, which soon, during the administration of President Bernardino Rivadavia (1826–1827), would become the *Departamento de Topografía* (Topographic Department) and then the *Departamento de Topografía y Estadística* (Department of Topography and Statistics). Spanish mathematician and surveyor Felipe Senillosa (1790–1858) was a key figure in these transitions (Zinny 1869, 23).

In the Chilean case, the project of a scientific and statistical exploration was entrusted to the French military man and *idéologue* Jean Joseph Dauxion-Lavaysse (1774–1829). His experience came from a research trip carried out to the Caribbean at the beginning of the nineteenth century, and which inspired the inquiries he would try to reproduce in Chile towards 1823 (Dauxion-Lavaysse 1820). Dauxion-Lavaysse developed his research applying first-hand data gathering, a combination of field notes and information obtained through questionnaires timely distributed among authorities and neighbours. By visiting each jurisdiction and interviewing its inhabitants, Dauxion-Lavaysse obtained direct evidence about climate, local history, economic activities, territorial demarcation, demographic conditions, and minutiae that enabled him to write the first drafts of his reports. Even though his mission was left unfinished (later critics accused him of superficial scientific training), several of his reports were published in Santiago newspapers between 1825 and 1826 (Lemmo 1967; Estefane 2017).

Even though these commissions were developed for government purposes, in the end they also fulfilled the need to produce images to attract the world's gaze. The judgement of foreigners was crucial for a generation of postcolonial intellectual and reformers trained in a cosmopolitan philosophy like the Enlightenment. However, there was no single position *vis-à-vis* their opinions. Some Creoles were hostile and reacted against the prejudices Europeans had about the continent, which most of the time, according to *El Telégrafo Mercantil*, derived from "the erroneous testimonies of quick and superficial scientific expeditions", referring to the testimonies generated by the famous scientific expeditions of the eighteenth century (*El Telégrafo Mercantil*, September 1, 1801, authors' translation). This prejudice was caricatured by Manuel

Lorenzo Vidaurre in his *Plan del Perú* (Peru's Plan). Here, he lamented that the entire population of the country was labelled lazy, negligent, and in constant apathy, because once a foreigner found some Peruvians sleeping at noon—doing the traditional *siesta* (napping) (Vidaurre 1823, 180).

In any case, this convergence of travellers and national portraits began to be built decades before independence, and it can be traced by looking at the inclusion of geographic reports written by foreigners in local journals and newspapers. *Mercurio Peruano* was one of the pioneers in this trend by publishing essays by the German naturalist Thadeus Häenke (1761–1816). *El Telégrafo Mercantil* in Buenos Aires followed the trend publishing articles by Häenke as well, particularly those on navigable rivers, but also included geographical studies and exploration reports by other authors (Navallo 2011). Another example of this was *El Araucano*, the Chilean official newspaper edited by Andrés Bello, who usually published translations of pieces from *The Quarterly Review*, *The Edinburgh Review*, and *The Proceedings of the Royal Geographic Society*. Most of these pieces dealt with South American geography but also with the development of statistics across the globe, in a clear attempt to reinforce the assimilation of this science in the country (Estefane 2017, 47, 83).

Another interesting case study on the links between foreign gaze and descriptive statistical research is Woodbine Parish (1796–1882), the British ambassador to Buenos Aires, who was also a renowned merchant, collector, and traveller. It is known that Parish had a strong influence on Rivadavia's decision to order the production of the first statistical account of the United Provinces of the Rio de La Plata, but he also wrote his own work, with similar intentions, entitled *Buenos Ayres and the Provinces of the Rio de la Plata*. For contemporary Europeans, works like these tended to fall under the broad and vague category of travel books simply because they featured news of remotes regions that ignite the imagination of audiences eager for adventures and exoticism. Instead of pleasing such type of readers, Parish offered a robust compilation of historical, geographical, statistical, and trade data. He even avoided including biographical information and wasting pages on irrelevant minutia. His goal was to present all the useful information he could accumulate and stuck firmly to that aim (Parish 1839).

The emphasis Parish imprinted on his work was not an expression of his genius. Books like his were part of a larger phenomenon: the presence of British economic interests in the region. When that fact is taken into account, it becomes clear that many of these publications were intended to provide information and data to future investors and traders in the metropolis. In a revealing twist, this connection reaffirmed the links between travel and political economy, in terms similar to those explored by Jean Baptiste Say, for whom travel could be seen as a kind of industrial experience (Say 1804). In fact, travel books written by foreigners reinforced this idea of a useful knowledge, legitimised by the credentials of scientific precision and by their display in the manner of quantitative statistics, that is, through tables and graphs. These interventions were very close to Carl Linnaeus' understanding, for whom the ideal traveller was a sort of 'industrial spy' (Koerner 1996, 152).

The progress of science and the circulation of useful knowledge were part of Parish's intellectual goals, but also part of its diplomatic mission. Indeed, his residency in the United Provinces had little to do with sociability and protocols, and a lot with geographic research, population measurement, identification of economic resources, the study of botany, zoology, etc. Although all these disciplines were different, it was common to see them converge in the statistical research programmes of the nineteenth century, which oriented the specialised investigation to the production of national inventories. What is interesting is that thanks to his diplomatic position Parish managed to articulate an efficient network of informants and collaborators who were crucial to his statistical programme. At some point he even admitted that, due to the difficulties in obtaining information from local authorities, he had to rely on two of his compatriots, who by then resided in the Argentine interior provinces: John Gillies, in Mendoza, and Joseph Redhead, who had lived for decades in Salta. These two "men of science" would have provided him with "variety of information" which he could never have obtained "from other sources" (Parish 1839, 35).

Parish also enthusiastically encouraged the exploration of the interior of the country. For example, in 1827 he commissioned John Gillies to travel to the south of Chile and across the Andes through the Las Damas and Planchón passes to explore areas not yet described, discover new crossing routes through the Andes, and find navigable rivers to connect these two regions. On this last excursion, Gillies crossed into Chilean territory and reached the cities of Curicó and Talca, where he explored the Maule River to evaluate its navigation capabilities (Parish 1839, 323 and 326).

Parish himself was an active explorer. He travelled to the Diamante River, on the southern limit of the Province of Mendoza, continuing south to the border with Chile, and through the Andes of Upper Peru in the company of Joseph Barclay Pentland (1797–1873), the private secretary of Charles M. Ricketts, the British consul in Lima (Parish 1839, 325). Pentland was an important link in Parish's network. Trained as a geologist under the supervision of the French eminence George Cuvier, Pentland was commissioned to prepare a vast report on the new republic of Bolivia. The purpose of this survey, as Consul Ricketts announced to Minister George Canning at the Foreign Office, was to obtain "a knowledge of its natural productions, which in fact be said to be unknown, as no scientific person has heretofore given any detailed description of them". It was also aimed to determine the geography of the region, map it, and examine the geology of the country, describe "the extent and capability of the mines, with reference to the advantage of machinery and the probability of the employment of British capital in working them", "the examination of vegetable and animal productions", and to collect "objects of interest for the British Museum" (Pentland 1827, 174–175). Interestingly, this expedition had been Canning's wish for a while because he wanted to find someone capable of conducting a similar geographical survey of the interior of Peru and Ecuador.

Actually, that commission had also been the wish of Simón Bolivar, who had been particularly eager for a scientist to tour Bolivia and write a report on the capacity of the mines. Coincidentally, Thomas Rowcroft, the unfortunate first British consul in Lima, had proposed the name of Joseph Gillies for this task, but the authorities thought otherwise and sent Pentland, who travelled the country in 1827. This was how Pentland wrote the first major description of Bolivia in an early period of its history, including a brief historical overview, geographical and statistical news, a report on the "present states of mines", and a brief on its political situation (Pentland 1827, 174–175).

Parish was an avid collector of useful data from the present, as well as of manuscripts with historical and geographic information gathered by colonial authorities and explorers. He found an important part of this material in old archives in Buenos Aires and in private collections of local researchers. He also compiled maps and topographic surveys of the Province of Buenos Aires that were expressly prepared for him by order of General Juan Manuel de Rosas, then governor of the capital. In the end, he assembled what he considered to be "a considerable collection of manuscripts, maps, and of inedited papers respecting countries of which the greater part of the world is, I believe, in almost absolute ignorance" (Parish 1823–1832).

Woodbine Parish did not create this collection out of an irrepressible antiquarian drive. While there may have been some of that, the main purpose was to gather useful knowledge and valuable data that could be used for the advancement of industry and commerce and for the promotion of British capital. It was a purpose not very different from that stated by the editors of the Chilean newspaper *El Progreso* in 1843:

> Not having knowledge about the age, status, and occupation of each individual, how can someone calculate the physical and moral strength, the total of the productive and consuming class, the daily drives and the wealth of the nation [...] Without reliable knowledge, taxes cannot be established according to the principle of justice and convenience that a proportional distribution prescribes; the industry in all its branches can be harmed instead of protected; foreign trade will not be able to be directed and promoted profitably; wandering around with no references, results would be the only factor on which to rely to judge the pertinence of the decisions taken. (*El Progreso*, November 17 and December 8, 1843, authors' translation)

This was how the more pragmatic vision of statistics became predominant, a vision less concerned with the old 'moral principle' and more focused on the precise measurement of what would allow contemporaries to know the resources and the natural wealth of their territories.

Conclusions

The overlapping meanings attributed to statistics during the first decades after independence expressed the most significant tensions regarding this type of knowledge in an environment of rapid institutional change. This review of testimonies by Creole intellectuals, politicians, and reformists, as well of foreign diplomats and scientists, show how the use of statistics as a framework for the definition of moral reform projects was displaced by a more pragmatic approach to this science. The new understanding was mainly driven by an increasing and decisive local identification with foreign interests, strongly related with the economic role that these new nations would begin to play on the global stage. Because of this transition, statistics ceased to be understood as an enlightened task entrusted to well-trained polymaths, or Creoles eager to show their loyalty to their homelands, to become a governmental task that could only be assumed by well-organised state offices, with research plans aimed at identifying available resources and measuring the productive potential of the territory.

In this way, a knowledge linked since the late eighteenth century to political economy, mostly focused on moral regeneration programmes that sought to reverse the prostration in which the Spanish administration had left the continent, or focused on disciplining and acclimating the principle of political representation, ended up displaced by the understanding of statistics as a descriptive tool that would allow to measure and display the richness and potential of these territories—and in dialogue with other sciences, such as geology or botany. Statistics began to operate as a knowledge that made these new nations 'legible' and 'consumable' to the eyes of the world. Thus, the replacement of the 'philosophical eye' by the 'industrial spy' is key to understanding the physiognomy of the future statistical offices that will emerge in the coming decades as well as to analysing how statistics became part of the definition of a new colonial pact with the global powers of the nineteenth century.

References

Bagú, Sergio. 1966. *El plan económico del grupo Rivadaviano (1811–1827). Su sentido y sus contradicciones, sus proyecciones sociales, sus enemigos*. Santa Fe: Universidad Nacional del Litoral. Instituto de Investigaciones Históricas.

Chile. 1823. *Constitución Política del Estado de Chile, promulgada en 29 de diciembre de 1823*. Santiago: Imprenta Nacional.

Dauxion-Lavaysse, Jean Joseph. 1820. *A Statistical, Commercial, and Political Description of Venezuela, Trinidad, Margarita and Tobago*. London: C. and W. B. Whittaker.

De Solano, Francisco, ed. 1988. *Cuestionarios para la formación de relaciones geográficas de Indias, siglos XVI/XIX*. Madrid: Consejo Superior de Investigaciones Científicas, Centro de Estudios Históricos, Departamento de Historia de América.

Destutt de Tracy, Antoine-Louis-Claude. 1824. *Tratado de economía política. Traducción al español por don M.V.M*. Vol. 2. Madrid: Librería de Rosas.

Egaña, Juan. 1813 [1887]. "Proyecto de Constitución para el Estado de Chile, compuesto por don Juan Egaña, miembro de la comisión nombrada con este objeto por el Congreso de 1811, y publicado en 1813 por orden de la Junta de Gobierno". In *Sesiones de los Cuerpos Legislativos de la República de Chile, 1811 to 1845*. Santiago: Imprenta Cervantes.

El Amigo de la Verdad. 1823. Santiago

El Araucano. 1835. Santiago.
El Censor. 1817. Buenos Aires
El Censor de la Revolución. 1820. Santiago.
El Curioso. 1821. Buenos Aires.
El Mercurio de Chile. 1822. Santiago.
El Progreso. 1843. Santiago
El Telégrafo. 1817. Santiago.
El Telégrafo Mercantil. 1801. Buenos Aires
El Verdadero Liberal. 1827. Santiago.
Estefane, Andrés. 2017. Elusive Numbers. State Knowledge and Bureaucratic Organization in Chile (1750–1911). PhD dissertation, State University of New York at Stony Brook. http://hdl.handle.net/11401/78299. Accessed 28 September 2020.
Gay, Claude. 1842 [1962]. "Letter by Claudio Gay to Manuel Montt, Valparaíso, June 23, 1842". In *Correspondencia de Claudio Gay. Recopilación, prólogo y notas de Guillermo Feliú Cruz and Carlos Stuardo Ortiz*. Santiago: Ediciones de la Biblioteca Nacional.
Gorriti, Juan Ignacio. 1836. *Reflecciones sobre las causas morales de las convulsiones interiores en los nuevos estados americanos y examen de los medios eficaces para reprimirlas*. Valparaíso: Imprenta del Mercurio.
Koerner, Lisbert. 1996. "Carl Linnaeus in his Time and Place". In *Cultures of Natural History*, edited by Nicholas Jardine, Anne Secord, and Emma C. Spary. Cambridge: Cambridge University Press.
La Abeja Argentina. 1822. Buenos Aires.
La Abeja Chilena. 1825. Santiago.
La Década Araucana. 1825. Santiago.
Larrea y Loredo, José de. 1875. Memoria dispuesta para la representación nacional por el señor Ministro de Estado en el Departamento de Hacienda. In *Documentos históricos del Perú colectados y arreglados por el coronel de caballería y fundador de la Independencia Manuel de Odriozola*, edited by Manuel de Odriozola. Vol. 7. Lima: Imprenta del Estado.
Lemmo, Angelina. 1967. *Dauxion Lavaysse y su obra*. Caracas: Universidad Central de Venezuela.
Mercurio Peruano. 1791. Lima.
Monteagudo, Bernardo. 1823. *Memoria sobre los principios políticos que seguí en la administración del Perú, y acontecimientos posteriores a mi separación*. Santiago: Imprenta Nacional.
Navallo, Tatiana. 2011. "La introducción a la Historia Natural de Tadeo Haenke y las representaciones de la naturaleza en la conformación del Alto Perú." *Antíteses* 4 (8): 703–728. https://doi.org/10.5433/1984-3356.2011v4n8p703.

Núñez, Ignacio. 1825. *Noticias históricas, políticas, y estadísticas de las Provincias Unidas del Río de La Plata: Con un apéndice sobre la usurpación de Montevideo por los gobiernos portugués y brasilero*. Londres: Publicado por R. Ackerman.

Otero, Hernán. 2006. *Estadística y nación. Una historia conceptual del pensamiento censal de la argentina moderna, 1869–1914*. Buenos Aires: Prometeo Libros.

Parish, Woodbine. 1839. *Buenos Ayres and the Provinces of the Rio de la Plata, their present state, trade, and debt*. London: John Murray.

———.1823–1832. *FO 354 Woodbine Parish Papers. Entry Books of Letters Received and Sent, Commercial Reports, etc, Belonging to Sir Woodbine Parish, Special Agent and Consul General at Buenos Aires from 1823 to 1832*. London: The National Archives in Kew.

Pentland, Joseph Barclay. 1827 [1974]. *Report on Bolivia*. London: Offices of the Royal Historical Society.

Quilter, Jeffrey, and Gary Urton, eds. 2002. *Narrative Threads. Accounting and Recounting in Andean Khipu*. Austin: University of Texas Press.

Randeraad, Nico. 2010. *States and Statistics in the Nineteenth Century: Europe by Numbers*. Manchester: Manchester University Press. https://doi.org/10.7765/9781526147530.

Salinas, Alejandro. 2001. *Estadísticas y contribuciones en el Perú, siglo XIX*. Lima: Universidad Nacional Mayor de San Marcos. Seminario de Historia Rural Andina.

Salomon, Frank. 2004. *The Cord Keepers. Khipus and Cultural Life in a Peruvian Village*. Durham and London: Duke University Press.

Say, Jean Baptiste. 1804. *Tratado de economía política o exposición sencilla del modo con que se forman, se distribuyen y se consumen las riquezas. Tomo 1*. Madrid: En la Oficina de Pedro María Caballero.

Spary, Emma C. 1996. "Political, Natural, and Bodily Economics". In *Cultures of Natural History*, edited by Nicholas Jardine, Anne Secord, and Emma C. Spary. Cambridge: Cambridge University Press.

Urízar Garfias, Fernando. 1835. *Repertorio chileno. Año de 1835*. Santiago: Imprenta Araucana.

Vidaurre, Manuel Lorenzo. 1823. *Plan del Perú, defectos del gobierno español antiguo, necesarias reformas. Obra escrita por el ciudadano Manuel de Vidaurre*. Philadelphia: Impreso por Juan Francisco Hurtel.

Westergaard, Harald. 1969. *Contributions to the History of Statistics*. New York: August M. Kelley Publishers.

Woolf, Stuart. 1989. "Statistics and the Modern State." *Comparative Studies in Society and History* 31 (3): 588–604. http://www.jstor.org/stable/178772.

Zinny, Antonio. 1869. *Efemeridografía argirometropolitana hasta la caída del gobierno de Rosas*. Buenos Aires: Impr. del Plata.

CHAPTER 3

Mapping Numbers: Statistics, Cartography, and the Making of National Space in Brazil

Alexandre de Paiva Rio Camargo

Introduction: Reasons for a Combined Study

Statistics and cartography are usually seen as academic knowledge produced by scientific societies and universities, as disciplines aimed at distinguishing different countries and enabling their comparison within a global space. However, more recent and still uncommon is the approach to their performative power as technologies for action on the realities they supposedly represent and which they help to create and establish under the pretext of only describing them. Maps always delimit space from a perspective of identity and difference, transforming what is heterogeneous and discontinuous into homogeneous and continuous spaces, either by fractionating previously constituted totalities or, conversely, by forging a new unity from fragments. Similarly, censuses and statistics establish

A. de P. R. Camargo (✉)
University Research Institute of Rio de Janeiro (IUPERJ), Candido Mendes University (UCAM), Rio de Janeiro, Brazil
e-mail: alexandre.camargo@pq.cnpq.br

C. T. Lanata-Briones et al. (eds.), *Socio-political Histories of Latin American Statistics*, Studies of the Americas,
https://doi.org/10.1007/978-3-030-87714-9_3

the *sequential display* of population phenomena—mortality, schooling, mobility, etc—a privileged and typically modern way of reversing space and time, which are then recreated, accumulated, and combined into devices such as tables, graphs, and diagrams (Latour 2000, 349–420).

As of the 1980s, the historiography of cartography and of statistics incorporated the constructivist lexicon of the social sciences, shedding light on the opacity behind numbers and maps' purported transparency, on what is hidden behind their displays, how they operate power relations, displacing and transforming people, groups, and places. Pioneer authors such as Brian Harley (1987) and Christian Jacob (1992) propose an approach to maps as artefacts, pointing to the different ways cartographic images are translated into cultural representations loaded with political messages, present in their explicit contents, their conventions, distortions, and decorations. The map is seen as a document that is both transparent—since it appears to contain the physical landscape and territory it represents—and opaque—for imposing its own textuality on them.

At the same time, the production of statistics came to be regarded as a practice relevant to the social sciences, not only because of its more obvious political effects. Sociologists as Alain Desrosières, epistemologists as Ian Hacking, and historians as Theodore Porter became interested in its unique features: the social relations involved in the chain of statistical production, the confidence, and the authority that these numbers usually inspire, the systems of classification, and representation of the world that they propose. This interest initially underlined the social and political uses of statistical sources, classifications, and instruments (INSEE 1987). Later, a research agenda would be formulated by Desrosières (2008, 7–20) focusing on restoring the meaning sought by a given statistical enterprise, which was sacrificed every time numbers were isolated from their conditions of production and (re)inserted into the historical series of censuses and surveys.[1] The aim here is to reverse the circuit that turns statistics into a black box, since its stabilisation presupposes the ignorance of its contingencies of origin.

The historiography of statistics and cartography evolved in a relatively independent manner, each one exploring the particularities of its

[1] This is a systematic procedure carried out by administrators, economists, and social scientists who increasingly generate data aggregates to support their analyses.

respective objects. Most of the time, this approach overlooked their interfaces and technological intersections. In one field, the map remained as the visual support for presenting numerical information, just like graphs and tables, without receiving distinct methodological and analytical treatment. In the other field, numbers mattered as textual and informative complement, in any case, subordinated to the structure of the map. Both approaches ignored that the cartographic display of population flows imposes another mathematical convention—a distinct graphic language is required to represent social relations spatially.

Few authors highlight this significant particularity that justifies a cross-cutting approach to numbers and maps. An exception is Gilles Palsky, who devotes much of his work to the emergence of statistical cartography and the making of its conventions (1991, 1996, 1999). For Palsky, the cartographic representation of human activities imposes a radical reform in the graphic language, which enables maps "to express facts of intensity related to areas with unequal dimensions, while traditional cartography was based on the linear quantification of fixed distances, latitudes, longitudes, and altitudes" (1996, 61–62, author's translation). From the viewpoint of cartographic work, it was a matter of creating new relations between images and numbers, by employing another syntax, which involved the invention of symbols and conventions to express the quantitative variations of social phenomena gradually defined by the human sciences throughout the nineteenth century.

The first modern statistical map, attributed to Baron Charles Dupin, was made in France in 1826. It illustrated primary education, considered a sign of the moral development of nations. This issue enabled Dupin to represent a fundamental disparity between northern and southern France, by employing an original graphic method—the choropleth—consisting of colourings applied to each area to produce an effect of gradation or hierarchy (Friendly and Palsky 2007, 240). In this case, departments were given shaded tones that referred to the ratio between the number of inhabitants and the number of children attending school, thus producing the famous polarisation between the enlightened north and the ignorant south, one of the marks of French nationality.

From then on, the marriage between statistics and cartography was increasingly argued by different individuals within European scientific societies. Beyond extending the knowledge and the comparability of the various realities around the globe, the attempt to build a synoptic representation of the world was at stake, which would allow to "apprehend the whole at a single glance". More than "a numerical science of social facts, appropriated by all disciplines, statistics would become a universal science by addressing each and every one through image" (Debluë 2018, 56).

Another pioneering effort at a combined analysis is Benedict Anderson's *Imagined Communities* (2008), which examines the role of censuses and maps in assembling the imperialist enterprise. Anderson shows how the census fills in the formal topography of the map through demographic triangulation, thus allowing the territoriality of the colonial state to be founded and the bureaucracy to be distributed, thereby remodelling pre-existing ethnic and racial boundaries. Paradoxically, by modifying the terrain where colonised populations lived, these technologies of visualisation produced an unexpected effect: a political convergence to imagine and soon establish a common space—which would prove crucial in national liberation movements (2008, 226–255).

Inspired by Anderson, other works focus on the role of censuses and maps in the construction of the state's authority and its power to name and make up people and places. From this angle, the history of statistics shows how official categories compete to divide groups within a population, separating and ordering them into codified hierarchies. Statistics not only establish these social divisions, but they also fix these categories of people, even when the margins of the groups are not clearly defined in social life (Alonso and Starr 1987). Counting the population to create positions of subjects and thus stabilise identities that are more real than others—traditional, familial, local—entailed attributing legal and symbolic effects to categories that have proved fundamental in the processes of ethnicisation of populations (Kertzer and Arel 2002). Categories such as "region" and "spoken language", for example, were invested in both statistical and cartographic forms. Such technological intersection reveals the performative power of censuses and maps to create new realities, new subjects, and spaces of government.

An example is Silvana Patriarca's (1996) seminal study on the role of statistics in the construction of the Italian territory as an organic entity to supplant the historical, cultural, ethnic, and linguistic diversity of the provinces, the major obstacle to the project of unification. Patriarca demonstrates how—between the 1840s and 1850s—the statistical discourse gradually lost its character as a descriptive device of administrative practices, incorporating new concepts and framings for depicting the peninsula, which constituted an alternative to the configuration of the pre-existing political order. The regional division of the territory enabled to focus on the economic and social problems shared by different provinces, which were understood, until then, in the classical terms of good or bad local government. In turn, through the spatial distribution of the inhabitants over each occupied area, the concept of population density replaced the primacy of absolute numbers on the amount of wealth and people as an indicator of a province's power, highlighting the variations within the territory and the need to optimise the relationship between population and resources (Patriarca 1996, 144, 153–155). Rather than ensuring the symbolic mediation of the Italian nation, quantification conferred a material basis to an organic and ordered whole, whose existence only lacked political recognition. Population density and regionalisation appear combined and related to a reality that is simultaneously invested by statistical and cartographic forms.

In the creation of the German nation state, Morgane Labbé identifies two modes of representation of German nationality and its borders. The first mode, visible on the maps of the 1840s, consisted in the creation of the political unity of the Prussian state, which would not be defined by its borders, but rather produced by the spatial distribution of the established linguistic groups, made visible by a simple cartographic procedure: the division of the territory according to the different languages spoken and the consequent attribution of their respective nationalities. The choropleth method was employed to emphasise that the existence of German groups did not coincide with the known divisions and borders, thus substantiating the right to dominance. Statistics appeared here alongside the historical record, projecting the precedence of each people in the occupation of the territory (Labbé 2010, 61–62). The second mode, introduced in the 1860s, converges with the rising of statistical cartography, conditioning the definition of nationalities not to the linguistic groups previously established in each state, but to their differential demographic growth in space, given by birth and immigration rates. In this new mode of representation, the territories at the limits of the Empire, populated by non-German groups, appear on the map based on their relationship to the German population, projecting a trend towards a demographic overlap, the new ground for the attribution of nationality and annexations. According to Labbé, "while early statisticians drew on history to illuminate the spatial distribution of nationalities, quantitative statisticians started from the present to search through statistical calculations and graphic methods for the factors of the future national settlement, giving rise to a new regime of action on population" (2010, 72, author's translation).

Following the path opened by the works mentioned above, this chapter contributes to the study of the technological intersections between censuses and maps. The cross-analysis proves to be especially fruitful in the investigation of the processes of national state-building, the centralisation of political authority, and the creation of new categories of people and spaces of government. In the first section, we examine the aspects that bring statistics and cartography together during the Brazilian Empire (1822–1889) and that point to a common status of maps and numbers as descriptive instruments of the sovereign power. The nineteenth-century tradition reveals the concern with coastal areas, transatlantic trade, border protection, indigenous tribes, natural resources, and the demonstration of a homogeneous population and a cohesive territory. Next, we address the political and cognitive conditions that lead to the expansion of the role of numbers and maps as a common ground for collective action, thus producing shared identities. By focusing on the spatial asymmetries of population, such as demographic gaps, illiteracy, and disease mortality, a visual representation of social interdependence takes place. In this sense, we analyse the emergence of statistical cartography and its rupture with the previous tradition. Therefore, we address two ways of governing and visualising population through numbers and maps, each one with its conditions of possibility. In transitioning from one to another, starting with the 1920 census, the emergence of the notion of 'physiographic regions' and the growing regionalisation of official data are worthy of special attention. Being a cartographic and statistical artefact at once, the region will be delimited as a new unit of analysis, rearranging the disposition and interpretation of numerical information on the national space, henceforth marked by the contrast between the backward north and the cosmopolitan south. Our analysis extends to the Vargas Era (1930–1945) and the first years of the Brazilian Institute of Geography and Statistics (*Instituto Brasileiro de Geografia e Estatística*, IBGE)—when territorial and population policies started to be framed by the regional division, and numbers began to point out problems according to the categories they helped to create. The extensive temporal scope of the chapter lies within the structural nature of the subject—the making of the national space—and the need to relate the state's long transformation to the institutionalisation of its two main political devices: statistics and cartography.

Maps and Censuses as Descriptive Instruments of Sovereign Power

Between Brazil's independence in 1822 and the consolidation of the imperial state in the 1850s, the term *pátria* (homeland) did not mean a society that recognised itself with its own identity and culture within a single space. According to Gladys Sabina Ribeiro, it was a synonym for the place exclusive to the 'good men', who recognised themselves through property and privilege bonds. Conveyed by words such as *pátria*, 'Portugal' and 'Brazil', their interests were passed off as those of everyone and as the common good. Thus, the word reached a limited community in which the object of loyalty and dignity was the king, now a constitutional emperor (Ribeiro 2008, 212). As a result, the idea of homeland was alternately used to denote both Brazil and the provinces—as suggested by Bernardo Pereira de Vasconcelos, who, despite being one of the most passionate opponents of separatism, referred to the province of Minas Gerais as "my homeland", in contrast to Brazil, which would be "the Empire". The distinction reveals that the concept of Brazil was a political construction, while the sentimental identification came from the province (Carvalho 2010, 77). In the early post-independence years, Brazil was an intellectual abstraction, a political modelling, and an absence in the collective imagination.

By means of an 'inward expansion', which Maria Odila Dias (1986, 160–184) calls "interiorisation of the metropolis", the concept of the Brazilian nation was forged to converge with the imperial state's foundations and self-image, materialised, among other strategies, "in the elaboration of a national language, literature, and history". The civilising project that the Empire embodied implied the affirmation of cultural autonomy, designed in varied initiatives: from the publication of magazines and programmatic texts to the creation of scientific societies and educational establishments. The Brazilian Historical and Geographical Institute (*Instituto Histórico e Geográfico Brasileiro*, IHGB), created in 1838, best embodied this role by producing a homogeneous vision of Brazil. In this vision, the political elite recognised itself as continuing the civilising task initiated by colonisation, linking nation, state, and crown into an indissoluble unity for the historiographical discussion regarding the country (Guimarães 1988, 6).

In this scenario, statistics and cartography enabled the visual demonstration of the crown's authority over a territory with a profoundly dispersed settlement, ranking the provinces according to their political weight and proximity to the Court. Maps, nominative lists of families, and population estimates also provided evidence for the arbitration of disputes over boundaries, avoiding the escalation of conflicts between provincial elites. The maps served to establish the civil, ecclesiastical, and judicial division of the provinces, to register the population, and to design the roads. The numbers reinforced the historical report and the oral tradition in order to prove the length of the occupation of a portion of the territory, thereby reinforcing the defence of its annexation or maintenance by a province.

Three aspects that bring statistics and cartography together in the nineteenth-century tradition are examined, pointing to a shared status of maps and numbers as descriptive instruments of sovereign power: (1) their modes of use; (2) the idea of population that they reproduce and make visible; and (3) their conditions of production and authentication of truth.

Statistics were the basis of electoral registration. Parish priests defined the voters by the number of dwellings (or domiciles), determining the political importance of each village. The 1824 decree established that parishes with 100 households would have one representative; parishes with 150 dwellings would have two; those with 250 would have three; and so on (Lyra 1979, 6). The count considered only dwellings, not individuals, and the country's population was estimated by the number of people believed to live there.

Besides electoral registration, the other fundamental use of statistics was in the definition of territorial boundaries. Documentation analysed by Julio Paixão Santos (2018, 83–85) reveals that the term 'statistics' was frequently employed to determine limits between provinces, cities, and towns. There was a semi-permanent "statistical commission" in the provincial legislative assemblies to issue deliberative opinions on the matter of border conflicts between municipalities in the jurisdiction. On a national scale, something similar was reproduced. Statistical commissions were often formed in the Chamber of Deputies, being responsible for receiving complaints from residents and public authorities regarding the definition of boundaries, the creation of and belonging to parishes or villages, the elevation of localities' status, the creation of laws on colonisation, and the increase in the number of deputies.

Chorographies were by far the prevailing genre of cartography in the Brazilian Empire. They constitute a dense description not only of the territory and its geographical features but also of the people and institutions that shaped the local life. The chorographic maps produced in the nineteenth century represent the relationship between painting and verbal description because they associate the image of the territory or province with the reading and arrangement of the natural and political elements in space (Bann 1994, 240). The combination of painting and description signalled what was to be shown and how the map should be read, emphasising the graphic representation of the administrative hierarchies that overlapped in the territory: municipal, judicial, ecclesiastical, and military boundaries.

For Ilmar Mattos (2005, 17–18), chorographies subordinated the individuality of each natural element, each people, tribe, custom, and production to the idea of a cohesive territory, which underpinned the Empire's affirmation. The issue of sovereignty was tied to the preservation of the vast territory, putting the definition of the relations between the provinces and Rio de Janeiro in the foreground, which would allow the constitution of an internal organisation as part of a whole. Thus, the chorographies expressed the knowledge of the provinces according to a centralising principle. The provinces did not matter for their local history or their internal elements, but for forming a mosaic linked to a state that sought to define its territory to promote administrative, fiscal, and political control over the population (Macedo 2007, 84).

This centralised integration project was carried out through a specific representation of the territory that served to the definition of limits and to the border policy of the Empire, relegating the distribution and characteristics of the population to the background. The indigenous villages were important due to concerns with their assimilation and the possibility of using them in the colonisation effort, revealing the centrality of the border policy.

Immigration would also fulfil this purpose by establishing colonies in areas with indigenous presence, on the margins of strategic routes, aimed to make a step back or to assimilate part of the villages. Before the disruption of the African labour force, immigration was practised in Brazil to attend the border policy, being encouraged in the lands of southern Brazil, where the rather rarefied character of the population hindered the security of the borders and the production of food to supply the military (Klug 2009, 206).

By analysing the limitations in the legibility of the Brazilian Empire—the simplification of the objects mapped for the administration of the state—Diego Bissigo identifies a pattern in the maps produced by different provinces. The indigenous areas, although signed, had no demarcation because it was expected that assimilation would render this distinction obsolete. Without demarcation, they became amorphous and open to the expansion of civilisation by the hands of the foreign immigrant: "indigenous groups are represented, however without a 'geographical body', being reduced to a point when settled, but lacking distinctions to indicate their difference, and therefore made invisible in several senses". In contrast, colonised areas become progressively delineated, "either because of the investment in on-site measurement and demarcation work or because it was a characteristic that one *wished to* show" (Bissigo 2020, 149, 151, author's translation).

The triangulation between frontiers, settlements, and immigration played a key role in the visual representation of the population in maps, at both the provincial and national levels. The consolidation of the Imperial State in the 1850s and 1860s created the stability needed for initiatives of greater scope and financial importance. The first national census, the production of the first general chart, and the publication of the first Brazilian geographic atlas were visual narratives that acted as powerful pedagogical resources for creating a specific image of the country.

Created as a work to be displayed at the 1876 International Exhibition in Philadelphia, the *Carta Geral do Império* (*General Chart of the Empire*) did not divide the provinces, presenting the country in its intended unity and territorial integrity. Although the publication indicated the existence of numerous indigenous groups, they were distributed in such a way as to coincide with the thin dotted line of the country's frontier. The choice of pointing out these groups conveyed the idea that their existence was known and mapped, when, in fact, their concentration on the border was a fiction created to defend the interests of the Empire. The placement of the indigenous peoples in the great empty spaces of the north and west reinforced their alienation from the nation, while at the same time ensured a line of defence for Brazil's sovereignty. In the areas closer to the Atlantic, these indications practically disappeared, as if they did not have any natives, so that the areas receiving immigrants and foreign capital could present themselves as civilised (Bissigo 2020, 292–293).

In turn, the *Atlas do Império do Brasil* (*Atlas of the Empire of Brazil*), published by Senator Candido Mendes de Almeida in 1868, was the first cartographic document with declared pedagogical purposes, being dedicated to the students of the Colegio Pedro II. Combining maps and estimates of his own on the size of the population, Almeida sought to create an image of the Brazilian space capable of valuing the limits and figures proposed by him.[2]

The *Atlas* had four maps of the whole Brazilian Empire. A physical map, with a section containing a pedagogical representation of the main geographical features and a table showing the bordering countries. A second map displays the electoral divisions of the Empire, with their districts, number of electoral colleges, parishes, and electors. The third map lists the dioceses that had bishoprics, informs their dates of creation, number of parishes, and the population of the respective provinces under the responsibility of each bishopric.

The fourth map represents the administrative division of Brazil, presenting a statistical table with the provinces, their capitals, extension, number of jurisdictions and municipalities, as well as the size of the population, total and per province (Fig. 3.1). Furthermore, this map contains a supplement with the main routes of the first years of Brazil's discovery. Thus, the image of an entirely Catholic Brazil, the convergence between Church and state, the precedence of the electoral and ecclesiastical division over the demographic reality, as well as the combination between the historical and statistical record in the presentation of a consolidated dominion over the territory stand out.

In each map, provinces are arranged in alternating colours, a resource that subjected the internal division to the harmony of the whole. The method of colouring is used with an aesthetic purpose, separating the Empire from neighbouring countries, which are represented with lighter lines and undifferentiated colour. The numbers on the provinces' area and populations appear as an informative complement.

The case involving this *Atlas* points to the limits that, at this time, surrounded the technological intersections between maps and statistics. The visualisation of quantitative information was reduced to the direct register on the chart's content without modifying its structure. Atlas, charts, and chorographies showed the administrative divisions of the

[2] The Atlas was used in several border negotiations. Its author, the jurist, historian, and geographer Candido Mendes de Almeida (1818–1881) was an eminent individual in the Conservative Party, deputy between 1843 and 1871 when he became a senator. As active member of the IHGB, he contributed to the organisation of civil law and was one of the great defenders of the separation between temporal and religious power (Grinberg 2002, 113–114).

Fig. 3.1 Atlas do Imperio do Brazil (1868) (*Source* Almeida [1868, 39])

country, based on tables and projections of the population that had no relation to space. Lacking the necessary conventions to quantitatively express social relations and human groupings, the map represents them through their qualities, naming and plotting them on its surface—as in the case of the indigenous tribes, distributed as an imaginary line of defence along the unpopulated border.

The 1872 census, Brazil's first general enquiry, also points to the same trends concerning population and sovereignty. The census sought to understand the impact of the casualties provoked by the recently ended War of the Triple Alliance (1864–1870) on the country's recruitment capacity. Additionally, to reduce the impact of the transition in the labour regime that began with the abolition of trafficking in 1850, the census aimed to determine the number and characteristics of the slave population, by comparison with the free population and the influx of immigrants.

As was the case with maps, these goals were aligned with the assertion of a homogeneous population, under the guardianship and the civilising action of the Empire. Concerning religion, the census assumed that all nationals and foreigners listed as 'Africans'—a group that indiscriminately included all slaves, as well as freed people not born in Brazil—were Catholic. Only white immigrants, presumably European, could be classified as 'non-Catholic', a category that comprised Protestants, Jews, and individuals with other religious beliefs. The census suggested a population that was 99.72% Catholic, virtually presenting Catholicism as a naturalised characteristic of Brazilians, precisely like in the *Atlas* of Candido Mendes. Indigenous marriages and the conjugal unions between slaves—officially single—that did not conform to the precepts of the Church were openly disregarded (Brasil 1877, 14).

Language was the second determining factor behind the representation of a homogeneous population. In a scenario increasingly marked by the use of language as a defining element of the nation, justifying territorial annexations and jurisdictional transfer of populations between states, the 1872 census disregarded the recommendations of the International Statistical Congress of St. Petersburg, leaving out language variations and proclaiming Portuguese as the language spoken by all nationals (Camargo 2018; Botelho 2005).

Besides mirroring the idea of a population dependent on the sovereign power as an indicator of the dominion of the state, census and map shared a deeply hierarchical view of the national space, placing the definition of the relations between the provinces and Rio de Janeiro—the base of the central power—at the forefront. Although there was no defined protocol, the arrangement of the census data followed the same order of the provincial maps established in Candido Mendes' *Atlas*. The information begins with Amazonas and continues along the coastline (Pará, Maranhão, Piauí) until Rio Grande do Sul. Then, from east to west, the so-called 'central' or hinterland provinces are listed: Minas Gerais, Goiás, and Mato Grosso.

This order is replicated in different volumes of the census results. Coast and borders are favoured to the detriment of the hinterland. The coast's prominence does not arise from the settling of its territory but from geopolitics that made borders' occupation strategic for the country's security. There, the Court occupies an intermediate position, acting as the axis of integration between the extremes. This visual record followed the same logic of cartography, expressing the knowledge of the provinces according to a centralising principle and the image of a cohesive territory on which the unity of the Empire intended to be based.

While numbers are reduced to an informative complement in the atlas and in the chorographies, maps do not represent visual support for the census data, which only appears in tables. Hence, censuses and statistics lack visual means to symbolically associate population and territory in the affirmation or construction of nationality. If in traditional cartography there is a direct relationship between the actual distance and the distance reduced in scale, how can demographic and economic quantities be translated into images? Recalling Palsky (1991, 451, author's translation), "the history of statistical cartography is first of all the history of the construction of a new code". It would be necessary to create on the map graphic forms that, without requiring reference in the visible space, would transpose the population movements of the statistical series to the map's system of positions, thus altering its semiology.

Third and last, mapping and population counts also converged from the point of view of their conditions of production and authentication of truth. In the absence of a legitimate monopoly on statistical and cartographic authority, both were carried out by private individuals, notable men, politicians of the Empire, and a sort of entrepreneur armed with a practical knowledge of the terrain to be surveyed. They self-financed their works, which were usually topographical, statistical, geographical, and ethnographic at the same time. Throughout most of the nineteenth century, the Brazilian state relied on their work to create a representation consistent with its control over the territory. Official and private statistics coexisted, as did official and private maps, made either by notable individuals or by public institutions who also produced the underlying information. The Ministry of Agriculture, Commerce and Public Works, for example, created the first *General Chart of the Empire*, while the General Board of Statistics (*Diretoria Geral de Estatística*, DGE-B) was behind the 1872 census.[3]

The state's accreditation was not a guarantee of authority nor of superiority over the private enterprise either for the census or a map. In turn, an author's renowned status was generally enough to attest to the validity of an estimate or chorography or its use as a document in the resolution of a dispute. The case of the *Atlas of the Empire of Brazil* is emblematic in this regard. With it, Candido Mendes not only established a number for the country's total population but also drew limits for borders still under dispute among the provinces, directly influencing the configuration of the Brazilian territorial framework. His authority was repeatedly called upon, for example, in the dispute opposing Piauí to Ceará, whose defences were based on the dividing line stipulated by Mendes (Bissigo 2020, 224–225).

Candido Mendes' estimate was one of 16 listed by Joaquim Norberto in his book *Investigações sobre os recenseamentos da população geral do Império e de cada província de per si tentados desde os tempos coloniais até hoje* (Silva 1986). The work compiled all the population calculations up to 1870, intending to provide support for the census that was to be carried out at that time. Norberto placed the estimates on the same level as the surveys of the provincial governments, placing them under the title "*censuses of the general population of the empire*", without attributing any superiority to the comprehensiveness of the census, from the point of view of its value of truth.

Even after the DGE-B's creation, a significant part of the imperial elite continued to resort to the tradition of estimates to contest the numbers of the first general census. The census was contrasted with population projections, based on indirect calculations made by the well-known names of the Empire (Camargo 2016, 180–187; 2018, 426). Rui Barbosa's

[3] The DGE-B, created in 1871, was the first office destined to coordinate the statistical services of the Empire. Its role was to take the census—including the first national population survey—to organise the annual tables of births, marriages, and deaths, and to formulate the plans for each branch of statistics of the Empire (IBGE 1951, 1). On the DGE-B, see Senra (2006, 2007).

opinion on the reform of primary education in 1882, for example, aimed to reveal the country's educational backwardness.[4] Written six years after the release of the census results, he used estimates even for the periods when official statistics were available, projecting the total number of inhabitants at 10,500,000 (Barbosa 1947, 14), without even mentioning the figure of 8,419,672 of the census. Paradoxically, Barbosa criticised the DGE-B figures not for their technical errors, which were not pointed out, but precisely for their official nature, which would automatically place them at the service of the elite's propaganda. The attack of a liberal like Barbosa resonated with the conservative cabinet of the Viscount of Rio Branco—the DGE-B's creator—and with the person responsible for the census, Senator Manoel Francisco Correa, another leader of that party, who was the head of the government during almost the entire period of DGE-B's activities (Camargo 2021a, 516–517).

Thus, the use and credibility of figures and maps were subject to political rivalries and the composition of the imperial cabinet, lacking the institutional structures necessary for their stability and endurance. The production of provincial charts and censuses was constantly discontinued, even when already in process, due to changes of provincial presidents, appointed by the emperor. Furthermore, the validity of an estimate or a chorography could be reinforced or contested, depending solely on the political affiliation of its author and the existing ministerial set-up, whether liberal or conservative (Camargo 2016; Bissigo 2020).

This scenario reflected the specificities of the Brazilian state, which resorted to liturgical services and distribution of privileges to impose order and make its presence viable in distant regions.[5] The 1872 census is a direct result of this arrangement, mirroring the dependence on ecclesiastical structure and the administrative apparatus financially controlled by the local elites. The census commissions were not remunerated—contradicting the terms of international conventions—and each was formed by a majority of five reputable individuals, knowledgeable about their places of origin, and by three bureaucrats, that is, the secular arm of the state. In addition, they had the authority to choose and supervise the census takers,

[4] One of the fathers of Brazilian legal thought, Rui Barbosa (1849–1923) was the sole author of the preliminary draft that would result in the basis of the 1891 Constitution, the second longest lasting to date (Camargo 2021a, 505).

[5] Liturgical services refer to the negative privileges extorted from privileged groups to satisfy the administrative needs of the political community.

distribute honorific rewards, and apply fines and punishments (IBGE 1951, 5–6), a prerogative of the police and judicial apparatus.

Statistics and maps, even when official, did not constitute a common reference. Their validity rested on the prestige of their authors who, in turn, were subject to the ever-moving alliances of the political arena. However, as instruments of sovereignty, they served the administration, allowing for fiscal control, military recruitment, and the construction of the country's political and territorial unity. For the provincial elites, they represented the centralised domination of a state reliant on private power to secure public order and carry out its major undertakings, such as Brazil's first census.

Statistical Cartography and the Regionalisation of Brazil

Once the Republic was proclaimed in 1889, the liturgical contributions would not disappear but would change form. The new constitution of 1891 favoured decentralisation and oligarchical domination. It provided the federated states, formerly provinces, with broad autonomy, giving them control over various issues, including education and health—except for higher education establishments and port hygiene. Thus, the federalism of the so-called *Early Republic* (1889–1930) was characterised by the control of local oligarchies over the executive and legislative branches.

The haste in conducting the 1890 census only a month and a half after the proclamation points to an evident attempt to legitimise the new political regime, which sought to circumvent its low popularity by appropriating from the Empire the image and symbols of guarantor of progress (Carvalho 1990). The alphabetical listing of the states broke with the Empire's standard for the visual display of data, which favoured the coast and the borders, to the detriment of the hinterland, making the Court the axis between the provinces. The choice for alphabetical ordering would seem more neutral considering the federation, dismissing the hierarchical representation (Camargo 2015, 82–84). Despite this difference, tables remained the only visual record of census information. Despite their longstanding use in Europe, the graphic technologies that allowed the physical representation of the territory to be subordinated to the variations in population flow remained absent. This was partly due to the failure of various attempts by the International Statistical Congresses to standardise the language of statistical maps (Palsky 1999).

A turning point was the commemorative bulletin of the 1908 National Exhibition, which marked the centenary of the opening of the ports of Brazil to Portugal's friendly nations. Produced by the DGE-B, the publication was the first to associate cartography and statistics through demographic triangulation. One of the maps reveals an unprecedented regional division of the country without explaining the criteria applied (Fig. 3.2). There was no legal or conceptual basis for the division. The map sectioned the territory, assuming the region as a self-evident datum, based on the traditional contrast between the coast and the western border, which had guided the spatial imagination since the Empire, as witnessed in the 1875 *General Chart*.

The first known map of Brazil's demographic density pioneered the concept of region to represent the nation taking the coast as the basis for regional division. The seven 'geographical zones' were detached to visually translate that fundamental opposition, by slicing the 'coast' (*littoral*) into four subregions that cover the Brazilian territory from the north to the south. The so-called central zone is not placed within the geographical centre of the country. It is a unit whose design took the hinterland of each state, from north to south, as opposed to its coastal portion. The 'far west' (*extremo-oeste*) appears as an isolated and independent region, distant from the nation, a visual effect obtained by its detachment from the subregions that constitute both the 'coast' and the 'centre'.

This demographic map appears as a hybrid experiment between two markedly distinct spatial rationalities. On the one hand, it explores the territory in zones, according to the unequal distribution of the population, in a pioneering way. On the other hand, adopting the region does not impose on the map another taxonomy between the whole and its parts. It rather bends to the logic of the political map, according to which borders can encompass an entire region and the whole is understood based on the opposition between the coast and the hinterland of each state. It praises the Atlantic and European connection of the coast, the historical, commercial, and cultural centre of Brazil, more urbanised and densely populated, from where the country's entire economic policy was organised. It emphasises the borders, which expressed the new American orientation of the republic, through the acquisition of territories, bilateral agreements with neighbouring countries, arbitration of conflicts in the region, and diplomatic reconciliation with the United States given the actions of the Baron of Rio Branco. The colouring of the map does

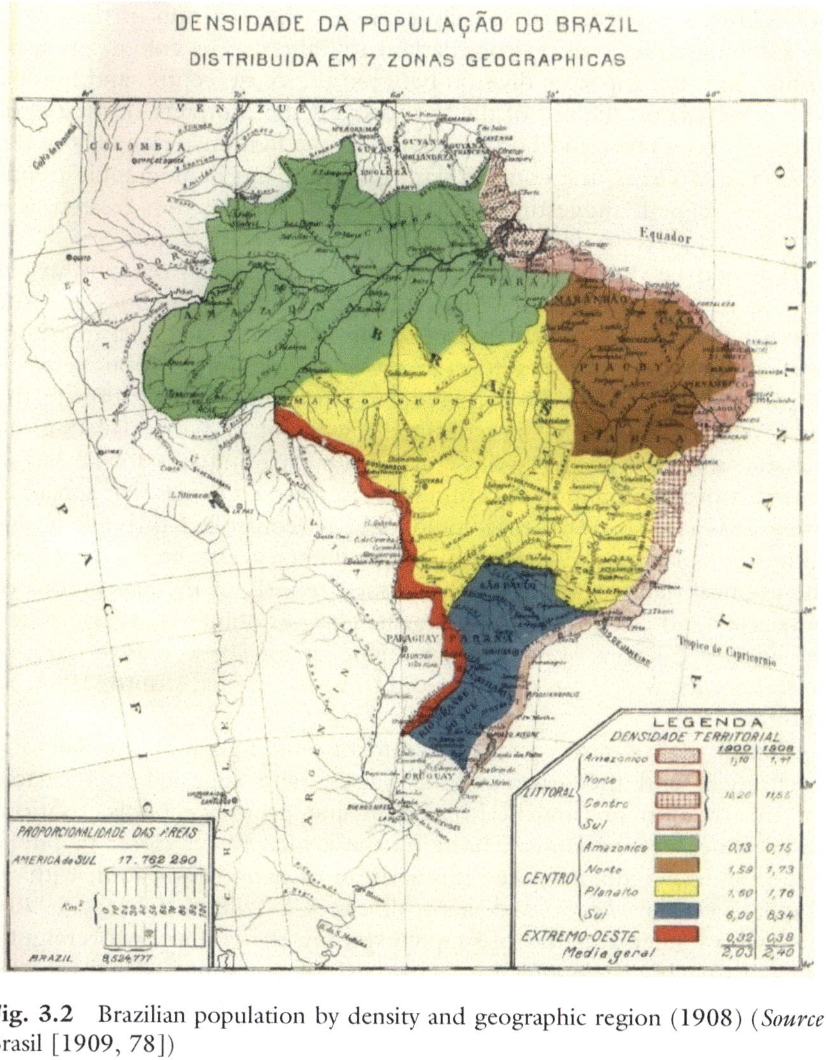

Fig. 3.2 Brazilian population by density and geographic region (1908) (*Source* Brasil [1909, 78])

not follow a gradient that visually expresses the variation in the country's demographic density, from darkest to lightest. The colours are used to highlight the tripartite division between the coast, centre, and border, with emphasis on the red of the 'far west'. If the centrality of the opposition between coast and border still recalls the figurative elements of the 1875 *Carta Geral*, the concern with precision and demographic trends becomes refined, suggesting that the map is not a fiction created as a propaganda piece to defend the interests of the sovereign power.

It would be only a few years later that statistical cartography would achieve its full political use. Merging international geopolitics and the internal dynamics of Brazilian society, the emerging 'national question' opened new opportunities for the technological intersection between maps and censuses. To be prepared for a potential participation in the First World War, a military enlistment was initiated that failed given the starvation of the lower classes. The war gave impetus to the health reform movement, aimed at battling diseases and endemics that spread in rural areas (Lima 1999) that could only be controlled through health education, placing the remarkably high illiteracy rate at the centre of the political debate. The widely published reports of the expeditions of physicians and scientists to the Brazilian *sertões* (hinterland) overstated the living conditions of the rural man (the *sertanejo*), seen as the root of *Brazilianness*, which was now believed to be threatened (Santos 1985; Sá 2009).

This picture of misery and abandonment caused great upheaval among intellectuals and the urban middle classes, who appealed to the state for the creation of rural health centres and primary schools. 'Nationalist leagues' were founded, bringing the armed forces closer to public health issues, combining the army's vision of national salvation with an emphasis on the *physical* and *moral* illness of Brazilians (Oliveira 1990). It was no longer a matter of simply exposing the country's sovereignty, but of managing the population as the basis of governmental activity. The political pact of oligarchic federalism was under pressure given the lack of cooperation between the union and the states in the areas of health and education, which soon after would be considered of public and national interest.[6]

[6] Camargo (2021b) analyses the role of vital statistics in the construction of public health as a national issue.

In this context, the statistical yearbook[7] and the 1920 census gave visual form and dimension to the issue of inequality between states, proving decisive in its establishment as a public and national problem. A rupture with the previous pattern of tabulation and presentation of census information took place. Demographic maps became frequent in official publications, dividing the territory according to each state's population density. In the yearbook and in the census, colouring began to follow the pattern of international conventions, employing a gradient of darker or lighter shades according to the recurrence of the phenomenon. The contrast between the densely occupied coastline and the hinterland—formed by large, depopulated states—emerged as a new image, both statistical and cartographic (Fig. 3.3). The concern with the border and the far west diminished in the structure of the map, modified to incorporate symbols illustrating the internal variations in the scale of growth. Colours, circles, and strokes were no longer aesthetic resources for exalting qualities or harmonising the cartographic whole. They became conventions that visually expressed the differences between demographic quantities, now in an internally unequal national space.

In the 1920 census, the region became a relevant unit for comparing population changes, discontinuing the political-administrative perspective adopted by the statistical discourse so far, which considered states the unit of analysis. The fact that regionalisation was applied precisely in the data on education and economic output is suggestive. In all cases, the Federal District emerged as a reference point, although without integrating the extremes or producing the image of a cohesive territory, as in the nineteenth-century tradition. The emphasis on the federal capital makes it possible to locate the shortcomings of the states and guide their future development.

In the section on education, the tabulation by regions is used twice: in monitoring school attendance and to portray the level of education of the population at school age (7–14 years old) and above (over 15 years old). Even though the general coefficient is considered extremely low for the country as a whole, great emphasis is placed on the asymmetries between the 'northern' and 'southern' states (Fig. 3.4).

Regionalisation exists also when tabulating agricultural production (Fig. 3.5). The census surveyed the number of rural establishments and

[7] There was only one Brazilian statistical yearbook produced during the Early Republic, covering the period 1908–1912.

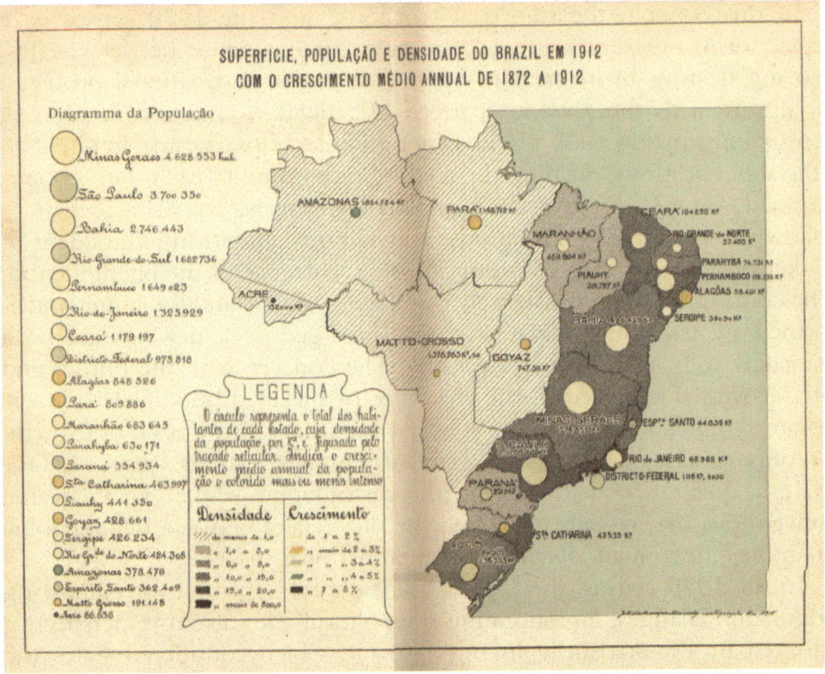

Fig. 3.3 Brazilian surface, population, and density in 1912, with average annual growth since 1872 (1916) (*Source* Brasil [1916, 252–253])

their cultivated area, crossing this data with the territorial extension of the states. The choice of the crossed variables highlighted that some states such as Pará and Mato Grosso could have the size comparable to an entire region without having a tenth of their cultivated area. The weak demographic densities of these states—in some cases absolute demographic voids—were also in evidence.

The regionalisation of the national space was not limited to the census volumes, producing significant repercussions in the press. The *Correio da Manhã*—one of the most important newspapers of the federal capital—published a series of articles on the number of agricultural establishments and their distribution across the country. One of them found that the number of rural establishments was not directly linked to the size or relevance of each state. It conveyed surprise when adding the territory of

Gráo de instrucção no grupo de 7 a 14 annos (*)

Zonas	HOMENS		MULHERES	
	Sabem lêr	Não sabem lêr	Sabem lêr	Não sabem lêr
Norte............	184.127 132	1.208.757 868	174.430 128	1.183.844 872
Sul...............	263.995 299	617.761 701	245.039 289	603.944 711
Centro............	123.147 163	634.660 837	106.084 147	615.644 853
Districto Federal	60.563 632	35.233 368	60.353 624	36.343 376
Espirito Santo....	9.978 198	40.458 802	8.187 164	41.601 836
Rio de Janeiro....	32.835 197	133.699 803	30.393 188	130.942 812
BRAZIL.....	674.645 202	2.670.568 798	624.486 193	2.612.318 807

Fig. 3.4 Education level of Brazilian population at school age (1920) (*Source* Brasil [1929, XVIII])

5 — Numero de estabelecimentos ruraes segundo a extensão territorial e a região

DESIGNAÇÃO DAS REGIÕES (2)	NUMERO DE ESTABELECIMENTOS			PORCENTAGEM DO NUMERO DE ESTABELECIMENTOS	
	Total	De menos de 101 hectares	De 101 e mais hectares	De menos de 101 hectares	De 101 e mais hectares
Estados maritimos do norte.............	188.930	127.355	61.575	67,4	32,6
" " " sul e Districto Federal..............	317.334	257.015	60.319	81,0	19,0
Estados centraes e Territorio do Acre...	141.889	79.509	62.380	56,0	44,0
TOTAL..............	648.153	463.879	184.274	71,6	28,4

Fig. 3.5 Rural establishments by territorial extension and geographic regions (1920) (*Source* Brasil [1923, XIV])

Mato Grosso and Goiás and finding that both encompassed an area larger than all the southern states together (today's south and southeast), and yet the total number of establishments in those two states did not reach 5% of the eight southern states. Compared to the northern states (present-day north and northeast), the share reached 43% (*Correio da Manhã* 1923). Findings like these seem self-evident today, and the contemporaries then certainly knew which states were the richest and which the poorest. The rupture taking place implied the discovery of the *measure* of inequality by an equivalence that sustained it. The region provided this equivalence, thus configuring a socio-spatial pattern of population phenomena, which the political division of the states contributed to isolate and mitigate.

Statistical regionalisation coincided with the awareness of social interdependence, given by the national movement that spread across the country, through the health reform in the *sertões* (Hochman 2012) and the campaigns to combat illiteracy and universalise primary education (Nagle 1997, 283–318). This movement propelled health and education as no longer simple branches of administration, but as spheres of the social world. Moreover, by measuring regional disparities, the census revealed a deeply heterogeneous and internally unequal national space. It provided inputs for the critique of the oligarchic order and for the centralisation of political authority, on which the census itself depended to broaden its reach as a common reference for social issues and policymaking (Camargo 2016, 358–377).

Alongside the awareness of social interdependence, the arrival of statistical cartography was in line with the development of another discursive field created by modernism. It was an artistic and intellectual movement which, from 1922 onwards, introduced the European avant-garde, though focusing on the elements of popular culture that would define the uniqueness of being Brazilian and protect its authenticity in a modern and cosmopolitan order. Through the efforts of modernist intellectuals to create an original literary universe and shape collective identities based on their visions of what popular culture would be (Ortiz 1985), nation and region appeared closely intertwined.

Under the stimulus of modernism, the 1920s saw the emergence of a new regionalism. Since the mid-nineteenth century, regionalism conceived the differences between the country's spaces as an immediate reflection of nature, environment, and race. Variations in climate, vegetation, and racial composition explained differences in customs, habits, and social and political practices. It was a naturalistic regionalism, in which Brazil appeared as a collection of landscapes with no synthesis or image-discursive structure to give unity to it. In contrast, modernism integrated regional elements into a national-popular aesthetic, rearranging them as signs in a new image and a new history for the country (Albuquerque Júnior 2011, 53, 68). The region began to be perceived as a social and cultural issue, projecting the moral personality of its characteristics and people into a geographical space.

The need to look for the roots and the feeling of national belonging became part of the debate concerning the three regional matrices that could give rise to the best society. The *Paulista*—from the state of São Paulo—regionalism addressed the *Bandeirante*[8] society as Brazil's gateway to democracy, with its supposedly innovative and anti-traditionalist characteristics: diffusion of polyculture and small property, diversity resulting from immigration and racial mixing, innovation, and entrepreneurship. The *Mineira* perspective—from Minas Gerais state—emphasised the psychosocial characteristics of the "*mineiridade*"—Minas Gerais' mode of being—its spirit of order and conciliation, manifested in sobriety, realism, and common sense as a way of life that resists time, ensuring a safe and stable passage from the traditional to the modern (Oliveira 1990, 195–196). The political speeches of the representatives of the northern states, previously dispersed, began to cluster around themes that affected national public opinion. The themes to be tackled—namely drought, messianism, and the *cangaço* (a sort of social banditry)—were the same that originated the very idea of the 'northeast'. The northeast was an area of power with borders serving as trenches to defend the endangered privileges of a decadent elite in search of preserving the past glory, the opulence of the *engenhos* (plantations), and the "peace and stability" of the Empire (Albuquerque Júnior 2011, 46–47). Northeastern regionalism not only invented the northeast. It also claimed the traditional values of patriarchal society, which should mediate the Brazilian transition to modernity.

The new regionalism cannot be reduced to initiatives and statements of a few individuals, nor to a mere cultural representation. It reconfigured existing power relations and constituted new subjectivities, promoting a deep cut-off in the narrative of identity and difference between the whole—nation—and its parts—regions. The work of inventing and naming people—as *Paulista*, *Mineiro*, or *Nordestino*—and 'regional' spaces was not only literary and cultural. To be embedded in the social imagination, these new categories of individuals and spaces needed to gain institutional forms, which would give them endurance

[8] *Bandeirantes* is the name given to explorers of the *sertões* in the colonial period who were mostly from São Paulo. In search of mineral and indigenous wealth for enslavement, they contributed to the territorial expansion of Brazil.

and stability. Statistics and cartography proved crucial for this, providing regional spaces with geographical reality and historical depth. They helped confirm the mythical time of the nation recognised in the map and conveyed sentiments and images compatible with the forms of belonging to regional spaces that operated new power relations.

The modernising agenda of the Vargas Era turned the regional theme into the starting point for the integration of the national whole. To confront the local oligarchies and the state powers representing them, the new regime invested in the creation of a national identity forged, entirely, on cultural and geographical regionalism. It strengthened the role of the municipality in public administration, eroding the wall of federalism above and below (Penha 1993, 108). Created in 1936, the IBGE—a national statistical office with much broader functions, resources, and autonomy than the DGE-B—was largely responsible for giving technical and conceptual consistency to the notion of region, to geographic regionalism, and to municipalism, in a project that aimed to build national political coordination. The IBGE played a decisive role in readjusting the Brazilian territorial framework.

The so-called geographic law of the *Estado Novo* (1938), which provided the country's territorial division, gave the IBGE a leading role in the territorial rearrangement. Different initiatives aimed to value the municipality as "the cell of national life" (Freitas 1943; IBGE 1945), not so much as to defend its autonomy but to undermine the prerogatives of the state oligarchies that dominated the Early Republic. In this sense, the IBGE standardised the boundaries and toponymy of the municipalities, which would give more accuracy to the census, reducing undercounting and duplicities. The statistical agency established the cartographic standardisation of municipal maps, enabling the production of the map of Brazil on the one-millionth scale. The office also concentrated the authority to arbitrate historical disputes of limits between different states, discussions that went back to the time of the Empire. The more accurate cartographic representation of the territory brought to light irregularities such as the improper demarcation of lands and the manipulation of borders and limits, illicitly conducted by the rural oligarchies (Davidovich 1985, 3).

Measures such as these directly involved the creation of territorial policies concerning their physical, economic, political, and population dimensions. The most grandiose was the exhibition of municipal maps, which included a ritual of burning state flags in the presence of the head of the nation.[9] These initiatives implied the control of the central government over the state to establish an order, a territorial homogeneity, built from the top down—through the region—and from the bottom up—through the municipalities. As the central body of technical elites, the IBGE gained prestige given the actions of President Getúlio Vargas. Its solid credibility helped to make political decisions emerge as the product of standardised analytical techniques, in a supposedly neutral language of intervention.[10] Some of the IBGE's main achievements, such as the setting of clear boundaries, the definition of contiguous and non-overlapping spaces, and the unification of nomenclatures and administrative circumscriptions, helped to spread and stabilise the state categories for the management of the population and the territory.

The geographical law of the *Estado Novo* also allowed the IBGE to have a division in every Brazilian municipality, turning it into a "capillary agency of central power" (Almeida 2000, 61, author's translation). On the one hand, the municipal statistical offices improved the network for data collection. Praised by the local authorities, the municipal agent should act as an evangelist to the population, educating them about the advantages of statistics, which undermined the remaining resistance to census surveys. On the other hand, the municipal agent directly advised the local administration. Playing the role of spokesperson for national policies, they also reported local demands to the upper strata of the hierarchy, bringing interests closer together and tailoring agreements (Senra 2008, 177–202).

The IBGE became the main operator of the municipalism advocated by the *Estado Novo*, according to which the municipality became organically and symbiotically intertwined with the central power (Camargo 2008, 37–48). Without mediations of territorial elites and political parties distorting this identity of ends, the articulation between the two levels would be backed by centralisation, bringing the two poles together. According to Marcus André Melo (1993, 88), the municipality

[9] A presidential decree issued by Vargas threatened the municipality that did not map its territory and send it to IBGE together with photographs of its geographical and urban characteristics within a year with the loss of political autonomy (Gomes 2010, 4).

[10] On several occasions, Vargas lent his political charisma to the IBGE, and associated it directly to his person, to strengthen territorial policies. This symbiosis took place since the IBGE's creation. The agency was placed within the Palácio do Catete, the official residence of the presidency.

emerged as a community sphere—therefore, pre-political—that shelters families and their values not yet distorted by political representation. This encounter of the central and the local established the public sphere over the private and particularistic order, in strong contrast to the liberal vision.

The IBGE played a key role in at least four initiatives in the making of the national space 'from above'. The first refers to the proposal for the territorial re-division of the Brazilian states, a much-debated subject in the 1930s and 1940s (Fig. 3.6). Conceived by Mario Augusto Teixeira

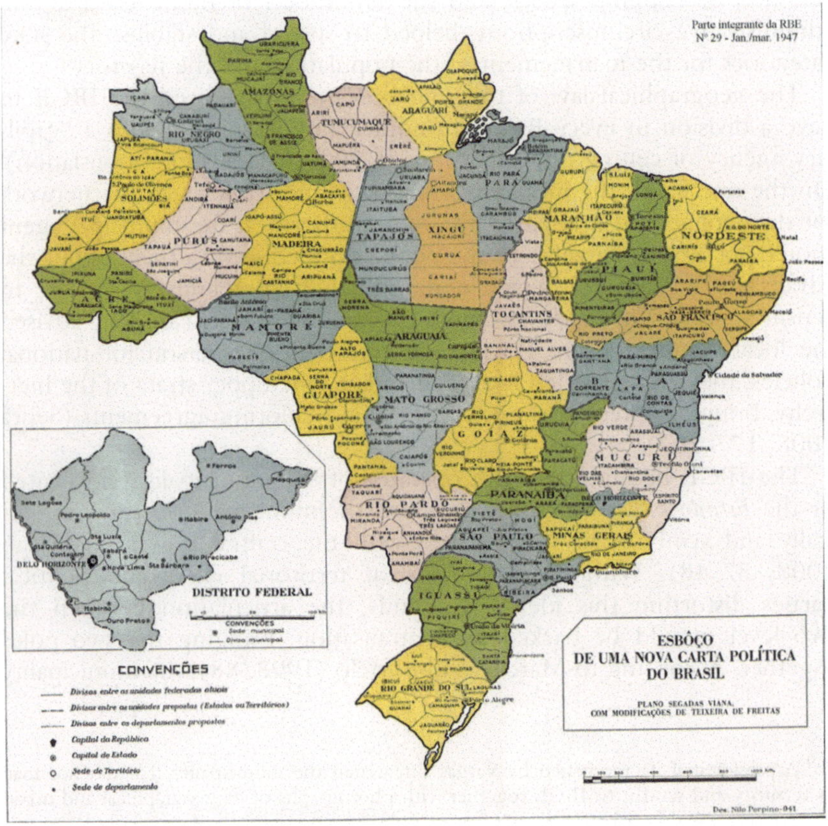

Fig. 3.6 Brazilian territorial re-division project, by Teixeira de Freitas (*Source* Freitas [1947])

de Freitas, the founder of IBGE,[11] the re-division followed the principles of territorial conformity and demographic equivalence among states (Freitas 1947, 699–714). In his opinion, imbalanced, depopulated, large, and powerful states would tend to separatism or disintegration. The territorial agenda would only cease to be a threat to Brazil when its units had area and political representation equivalent and suitable to their population. In Teixeira de Freitas' plan, no state would cede territory to another existing state and none would be suppressed. Small states would be merged into autonomous departments within a larger state. When adopted in sparsely populated regions, the new division would generate new states or territories equally divided into departments. Although it was not accomplished, the proposal shared the cognitive logic of statistical cartography by focusing on demographic and regional asymmetries in the population density and the distribution of resources to forge a federal balance among the states.

Second and most relevant was the regional division of Brazil, which established as physiographic zones the 'great regions' known today—north, northeast, south, east (present-day southeast), and centre-west (Fig. 3.7). Established in 1941 in the *Brazilian Journal of Geography* (*Revista Brasileira de Geografia*), published by the IBGE, the regional division developed by Fabio Macedo Soares Guimarães established a "single division" of the territory. The author defended the concept of a "great homogeneous region", arguing that this scale would allow regions to be established according to their logic "defined by nature"—which demonstrated to be "more suitable for didactic and educational purposes" and "more stable for the collection and comparison of statistical data on the national space" (Guimarães 1941, 368–369, author's translation). The division became official by decree on 31 January 1942, obliging all official statistics of states and municipalities to be tabulated according to Brazilian regional units (Valverde 2006, 271).

The third line of action, directly linked to the "didactic and educational purposes" of the official regional division, implied the development of a *pedagogy of space*, popularising the perspective of the state bureaucracy. This pedagogy was implemented in IBGE's courses, regularly offered

[11] Teixeira de Freitas was initially a delegate of the 1920 census in Minas Gerais, to then become head of that state's statistical office. With the revolution of 1930 and the rise of Vargas, he became head of the statistical services of the Ministry of Education to then design the IBGE. On Teixeira de Freitas, see Senra (2008) and Camargo (2010).

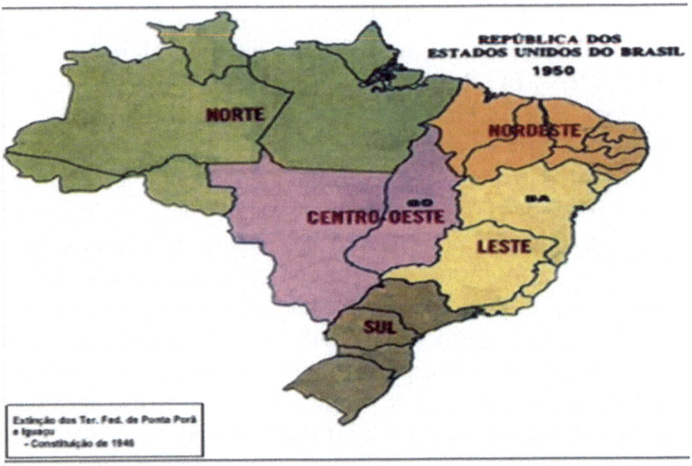

Fig. 3.7 Brazilian regions established in 1942 (1950) (*Source* IBGE [1995])

to secondary school teachers, in the geography manuals produced by its technicians, in its collections of wall maps widely distributed to the schools, and in several articles published in the *Brazilian Journal of Geography* that focused on the teaching of geography and the didactic use of cartography in the classroom (Camargo 2009, 27–30). These endeavours were combined to standardise the teaching practice and frame the approach of the regional perspective. Despite being already known, the conversion of the map into a logo was starting to become *systematic*—as it happens when the image of the perimeters of the geographical space is used as a representation of the nation and the nationality, amplifying its power of mobilisation (Anderson 2008, 241–242).

Last but not least, the regional division established a legal and stable framework for the regionalisation of statistical data, thus redefining its uses. The 1920 census had limited itself to regionalising data on school attendance and the cultivated area of rural establishments, but even then, just through commentary. There was no univocal criterion for the regionalisation of data, nor an official standard for representing the nation on a regional basis. Heterogeneous and uncertified regional divisions coexisted for the purposes of state action. In contrast, the results of subsequent censuses would be presented by regions, covering an increasingly larger set of themes, organised in the structure of the publication, and integrated into statistical series. Backed by the new legal landmark, the 1940 census regionalised the data on education, social security, unionisation, housing, colour, religion, activity by sex and age (IBGE 1950a, 71–167), cost of living, wage, and rural establishments (IBGE 1950b, 40–95, 298–420). Hence, the census became an instrument of wage indexation and of monitoring social rights, based on the regional differences created by the census itself.

The 1950 census went further by introducing regional comparison in the historical series of demographic surveys, using the official definition adopted in 1942 and going back to the 1872 census, by aggregating the provinces corresponding to each of the major regions. Alongside geographical regionalism, the statistical series shaped the historical depth of the region by calculating the differentials of population growth, fertility, education, occupation, production, and cost of living, and associating them to regional realities. Later, when statistics addressed the processes of internal migration, mobility, and social stratification, the 1970 census revealed the measure of income concentration, giving rise to a great controversy on poverty as an obstacle to the country's development (Malta 2010). Once again, the statistical visualisation of regional inequality fostered the rise of public debate regarding the future of Brazil given the divorce between economic growth and social development.

As it happened before, and would happen many times after, the region did not count only as a unit of analysis or a descriptive data of the census. At a deeper and more performative level, it served as a reference for the statistical conception of normality as recurrence, in which the discrepant values found by the census are classified according to the national average and the consequent distribution of the normal curve, highlighting the severity in the worst-performing states and regions. On the one hand, statistical and cartographic work solidified the region as a physical reality, inscribed in nature and prior to history itself. On the other hand, the naturalisation of the region rendered invisible the conventions that underpinned the production of numbers, increasing their opacity and resistance to criticism, converting them into pillars for state routines and collective action.

Shaped by statistical regionalisation and by geographical and cultural regionalism, the region became the foundation of the imagined community, a key aspect for the identity and alterity within the national space. Reproduced in textbooks, newspapers, television, infographics, and digital media and converted into a globally recognised symbol in the products of the cultural industry—often in the form of a map-logo—the region gained such solidity and stability that it is now no longer possible to imagine *Brazilianness* without it. At the same time, the perception of poverty, violence, and different forms of inequality—of income, race, gender—relies on the region, one of the main grounds for the knowledge and management of the population.

Final Remarks

This chapter analyses the Brazilian experience to demonstrate the heuristic possibilities of the cross-cutting approach of numbers and maps, focusing on the mutually conditioned relations between their technological changes and their modes of use, and the state-building process. From a methodological point of view, maps are not a mere visual support for the presentation of statistics, which requires a more thorough analysis of cartographic language. If their signs and conventions reduce numbers to an informative complement subordinated to the structure of the map—as in the *Atlas* of 1868 and the *Carta* of 1875—we are faced with a descriptive instrument intended to enlighten the sovereign power, indistinctly signed by an author or an institution, given the absence of a legitimate monopoly of statistical and cartographic production.

In this scenario, censuses and maps serve as legal proof in disputes, and they display a supposedly homogeneous population and organise a cohesive and hierarchical territory according to a centralising principle. State certification does not constitute authority or superiority of the census and the map over indirect estimates and chorographies made by notable individuals. In turn, political rivalries and the alternation of parties may suffice to discredit well-known surveys and discontinue initiatives in progress. In other words, censuses and maps, whether provincial or general, even when official, do not constitute common references because they *do not resist criticism*.

If, on the contrary, signs and conventions allow the representation of facts of different intensity in areas with unequal dimensions, as in the demographic map, a new syntax is established in the relationships between images and numbers. Instead of the traditional concern with borders, villages, and immigrant colonies, supported by a historical and static record focused on the domain of the territory, we find the dynamic record of population flows and the monitoring of their internal variations. This is not merely a new convention, since the technological change reflects and, at *the same time*, reinforces the movement of political coordination which, in Brazil, had a turning point in the 1920s, reaching its peak with the expansion of the state during the Vargas Era. Instead of a simple addition of the number to the previous content of the map, as in the nineteenth century, the two devices combine in a new political and cognitive logic, which multiplies their performative power.

Therefore, the emergence of statistical cartography in countries like Brazil should not be reduced to the circulation of scientific knowledge and the adherence to the recommendations of international forums, which foresaw the choropleth method at least since the International Statistical Congresses. It is also important to acknowledge the social conditions behind the emergence of biopolitics (Foucault 2009), a new regime of action on the population, focused on understanding the asymmetries and obstacles that prevent the development of its inner forces. The new regime implied action-oriented knowledge, which modified the descriptive approach of the statistical tradition to calculate the future factors of the national settlement, anticipating the field of possibilities that should guide governmental activity.

Under this new regime, the formation of an internally unequal national space took place in the different attempts to regionalise the country, combining statistical and cartographic forms, until the legalisation of the regional division, under the expertise of the IBGE. Our analysis of some of the institute's initiatives, in particular the stabilisation of state nomenclatures, the capillarisation of central power, and its decisive role in naturalising the region, shows that statistics and maps depend on a certain level of infrastructural power to become reliable references. In turn, masking the region's conventional basis increases the resistance of these devices to criticism, transforming them into pillars for collective action. The authority of numbers feeds back from the state's authority.

The technological intersections of numbers and maps can either be studied as an independent object or provide a framework for investigating the state's transformations and its relationships with society in the processes of historical and social change. The latter was our choice in this work. Following Desrosières—for whom quantification is the synthesis of two moments, that of creating a convention (*convenir*) and that of measuring (*mesurer*)—to quantify is also to visualise and imagine demographic and economic quantities in a coherent and unitary spatiality once it is stabilised by conventions of equivalence. This understanding requires further disciplinary and transatlantic dialogues.

References

Albuquerque Junior, Durval Muniz de. 2011. *A invenção do nordeste e outras artes*. São Paulo: Cortez.

Almeida, Candido Mendes de. 1868. *Atlas do Imperio do Brasil*. Rio de Janeiro: Lithographia do Instituto Philomathico.

Almeida, Roberto Schmidt de. 2000. "A geografia e os geógrafos do IBGE no período 1938–1998." PhD dissertation, Universidade Federal do Rio de Janeiro. https://biblioteca.ibge.gov.br/visualizacao/livros/liv66453.pdf. Accessed 12 May 2021.

Alonso, William, and Paul Starr, eds. 1987. *The Politics of Numbers*. New York: Russell Sage Foundation.

Anderson, Benedict. 2008. *Comunidades imaginadas: reflexões sobre a origem e a difusão do nacionalismo*. 2a. ed. São Paulo: Companhia das Letras.

Bann, Stephen. 1994. *As invenções da História: ensaios sobre a representação do passado*. São Paulo: Editora Unesp.

Barbosa, Rui. 1947. "Reforma do ensino primário e várias instituições complementares de instrução pública." In *Obras completas de Rui Barbosa*, 10. Rio de Janeiro: Ministério da Educação e da Saúde.

Bissigo, Diego Nones. 2020. "Um Estado de visão: a cartografia na legibilidade do Brasil Império." PhD dissertation, Universidade Federal de Santa Catarina. https://repositorio.ufsc.br/handle/123456789/220534. Accessed 12 May 2021.

Botelho, Tarcísio Rodrigues. 2005. "Censos e construção nacional no Brasil Imperial." *Tempo social* 17 (1): 321–341.

Brasil. 1877. Ministério dos Negócios do Império. *Relatório da Diretoria Geral de Estatística, dirigido ao Ministro dos Negócios do Império Dr. José Bento da Cunha e Figueiredo, por Manoel Francisco Correia, Diretor Geral em 31 de dezembro de 1876*. Rio de Janeiro: Tipografia de Hyppolito José Pinto.

———. 1909. Diretoria Geral de Estatística. *Boletim comemorativo da exposição nacional de 1908*. Rio de Janeiro: Tipografia da Estatística.

———. 1916. Diretoria Geral de Estatística. *Annuario Estatístico do Brasil. 1908–1912*. Rio de Janeiro: Tipografia da Estatística.

———. 1923. Diretoria Geral de Estatística. *Recenseamento do Brasil, realizado em 1 de setembro de 1920. Superfície territorial, área e valor dos imóveis recenseados. Categoria e nacionalidade dos proprietários. Sistema de exploração rural. Recenseamento do gado*. Rio de Janeiro: Tipografia da Estatística.

———. 1929. Diretoria Geral de Estatística. *Recenseamento do Brasil, realizado em 1 de setembro de 1920. População por estados e municípios, segundo o grau de instrução, por idade, sexo e nacionalidade*. Rio de Janeiro: Tipografia da Estatística.

Camargo, Alexandre de Paiva Rio. 2008. "Municipalismo e ruralismo: o IBGE e a organização nacional na era Vargas." In *O IBGE na história do municipalismo e sua atuação nos municípios*, edited by Nelson de Castro Senra. Rio de Janeiro: IBGE. https://biblioteca.ibge.gov.br/visualizacao/livros/liv38889.pdf. Accessed 12 May 2021.

———. 2009. "A Revista Brasileira de Geografia e a organização do campo geográfico no Brasil (1939–1980)." *Revista Brasileira de História da Ciência* 2 (1): 23–39. https://www.sbhc.org.br/arquivo/download?ID_ARQUIVO=42. Accessed 12 May 2021.

———. 2010. "Povoar o hinterland: o ensino rural como fronteira entre estatística e educação na trajetória de Teixeira de Freitas." *Revista Brasileira de História da Educação* 10 (2): 97–132. https://periodicos.uem.br/ojs/index.php/rbhe/article/view/38531. Accessed 12 May 2021.

———. 2015. "Dimensões da nação: uma análise do discurso estatístico da diretoria geral de estatística (1872–1930)." *Revista Brasileira de Ciencias Sociais* 30: 79–97. https://doi.org/10.17666/308779-95/2015.

———. 2016. "A construção da medida comum: estatística e política de população no Império e na Primeira República." PhD dissertation, Universidade do Estado do Rio de Janeiro.

———. 2018. "O censo de 1872 e a utopia estatística do Brasil Imperial." *Revista de História da Unisinos* 22 (3): 414–428. http://revistas.unisinos.br/index.php/historia/article/view/15474. Accessed 8 January 2021.

———. 2021a. "Para uma arqueologia da verdade estatística: reflexões a partir do projeto de reforma educacional de Rui Barbosa (1882)." *Revista Brasileira de História e Ciências Sociais* 13 (25): 500–529. https://doi.org/10.14295/rbhcs.v13i25.11938.

———. 2021b. "Estatísticas sanitárias e interdependência social na Primeira República." *História, Ciências, Saúde – Manguinhos* 28 (4): 1007–1035. https://doi.org/10.1590/S0104-59702021000400006.

Carvalho, José Murilo de. 1990. *A formação das almas: o imaginário da República no Brasil*. São Paulo: Companhia das Letras.

———. 2010. *Cidadania no Brasil: o longo caminho*. 13th ed. Rio de Janeiro: Civilização Brasileira.

Correio da Manhã. 1923. "Notas econômicas." January 16.

Davidovich, Fanny. 1985. *Um informe sobre a posição da geografia no IBGE*. Rio de Janeiro: IBGE.

Debluë, Claire-Lise. 2018. "De l'image numérique à la photographie: exposer les savoirs statistiques au tournant du 20e siècle." *Transbordeur: photographie, histoire, société* 2: 54–65.

Desrosières, Alain. 2008. *Pour une sociologie historique de la quantification*. Paris: Presses de l'École des Mines. https://doi.org/10.4000/books.pressesmines.901.
Dias, Maria Odila da Silva. 1986. "A interiorização da metrópole (1808–1853)." In *1822: dimensões*, edited by Carlos Guilherme Motta. São Paulo: Perspectiva.
Foucault, Michel. 2009. *O nascimento da biopolítica*. São Paulo: Martins Fontes.
Freitas, Mario Augusto Teixeira de. 1943. "O problema do município no Brasil atual." *Revista Brasileira de Estatística* 4 (16): 699–714. https://biblioteca.ibge.gov.br/visualizacao/periodicos/111/rbe_1943_v4_n16.pdf. Accessed 19 May 2021.
———. 1947. "O reajustamento territorial do quadro político do Brasil." *Revista Brasileira de Estatística* 8 (29): 109–124. https://biblioteca.ibge.gov.br/visualizacao/periodicos/111/rbe_1947_v8_n29.pdf. Accessed 19 May 2021.
Friendly, Michel, and Gilles Palsky. 2007. "Thematic Maps and Diagrams. Visualizing Nature and Society." In *Maps: Finding Our Place in the World*, edited by James Akerman and Robert Karrow. Chicago: University of Chicago Press.
Gomes, Maria do Carmo Andrade. 2010. "A exposição nacional dos mapas municipais: a encenação nacionalista da imagem cartográfica." In *Anais do Simpósio Iberamericano de História da Cartografia*, 3. São Paulo: Edusp. https://3siahc.files.wordpress.com/2010/05/exposicao-nacional-de-mapas-municipais1.pdf. Accessed 28 February 2021.
Grinberg, Keila. 2002. "Candido Mendes de Almeida." In *Dicionário do Brasil Imperial (1822–1889)*, edited by Ronaldo Vainfas. Rio de Janeiro: Objetiva.
Guimarães, Fábio de Macedo Soares. 1941. "Divisão Regional do Brasil." *Revista Brasileira de Geografia* 3 (2): 318–373.
Guimarães, Manoel Luís Salgado. 1988. "Nação e civilização nos trópicos: o Instituto Histórico e Geográfico Brasileiro e o projeto de uma História Nacional." *Estudos Históricos* 1 (1): 5–27. http://bibliotecadigital.fgv.br/ojs/index.php/reh/article/view/1935. Accessed 19 May 2021.
Harley, Brian. 1987. *The New Nature of Maps: Essays on the History of Cartography*. Baltimore and London: John Hopkins University.
Hochman, Gilberto. 2012. *A era do saneamento: as bases da política de saúde pública no Brasil*. 3a ed. São Paulo: Hucitec.
IBGE. s.d. *Recenseamento do Brazil em 1872*. Rio de Janeiro: Typographia G. Leuzinger.
———. 1945. *Problemas de base do Brasil*. Rio de Janeiro: IBGE.
———. 1950a. *Recenseamento geral do Brasil realizado em 1 de setembro de 1940: Censo demográfico. População e habitação*. Rio de Janeiro: IBGE.
———. 1950b. *Recenseamento geral do Brasil realizado em 1 de setembro de 1940: Censos econômicos. Agrícola, industrial, comercial e dos serviços*. Rio de Janeiro: IBGE.

———. 1951. *Legislação básica dos recenseamentos de 1872 e 1890*. Rio de Janeiro: IBGE.

———. 1995. *Anuário estatístico do Brasil*. Rio de Janeiro: IBGE.

INSEE. 1987. *Pour une histoire de la statistique*. Paris: Económica.

Jacob, Christian. 1992. *L'Empire des cartes: approche théorique de la cartographie à travers l'histoire*. Paris: Albin Michel.

Kertzer, David, and Dominique Arel, eds. 2002. *Census and Identity: The Politics of Race, Ethnicity and Language in National Censuses*. New York: Cambridge University Press.

Klug, João. 2009. "Imigração no sul do Brasil." In *O Brasil Imperial, volume III: 1870–1889*, edited by Keila Grinberg and Ricardo Salles. Rio de Janeiro: Civilização Brasileira.

Labbé, Morgane. 2010. "Les frontières de la nation allemande dans l'espace de la carte, du tableau statistique et de la narration." In *Les espaces de l'Allemagne au XIXe siècle: frontières, centres et question nationale*, edited by Catherine Maurer. Strasbourg: Presses Universitaires de Strasbourg.

Latour, Bruno. 2000. *Ciência em ação: como seguir cientistas e engenheiros sociedade afora*. São Paulo: Editora Unesp.

Lima, Nísia Trindade. 1999. *Um sertão chamado Brasil: intelectuais e representação geográfica da identidade nacional*. Rio de Janeiro: Iuperj/Revan.

Lyra, Augusto Tavares de. 1979. *Instituições políticas do Império*. Brasília: Senado Federal; Universidade de Brasília. Coleção Bernardo Pereira de Vasconcellos, 16.

Macedo, Adriana Mattos Clen. 2007. "As corografias e a cultura histórica oitocentista." PhD dissertation, Universidade Federal do Rio de Janeiro.

Malta, Maria de Mello. 2010. "A controvérsia da distribuição de renda no pensamento econômico brasileiro: a ampliação do debate teórico sob a conjuntura dos anos 1970." XXXVIII Encontro Nacional da ANPEC. Salvador.

Mattos, Ilmar Rohloff de. 2005. "Construtores e herdeiros: a trama dos interesses na construção da unidade política." *Almanack Braziliense* 1: 8–26. https://www.revistas.usp.br/alb/article/view/11601. Accessed 19 May 2021.

Melo, Marcus André. 1993. "Municipalismo, *nation building* e a modernização do Estado no Brasil." *Revista Brasileira de Ciências Sociais* 6 (23): 85–99. http://www.anpocs.com/images/stories/RBCS/23/rbcs23_07.pdf. Accessed 19 May 2021.

Nagle, Jorge. 1997. "A educação na Primeira República." In *História geral da civilização brasileira: o Brasil republicano, sociedade e instituições (1889–1930)*, edited by Boris Fausto. Rio de Janeiro: Bertrand Brasil.

Oliveira, Lucia Lippi. 1990. *A questão nacional na Primeira República*. São Paulo: Brasiliense.

Ortiz, Renato, 1985. *Cultura brasileira e identidade nacional.* São Paulo: Brasiliense.

Palsky, Gilles. 1991. "La cartographie statistique de la population au XIXè siècle." *Espace, populations, sociétés* 9 (3): 451–458. https://doi.org/10.3406/espos.1991.1487.

———. 1996. *Des chiffres et des cartes: naissance et développement de la cartographie quantitative française au XIXe siècle.* Paris: Comité des travaux historiques et scientifiques.

———. 1999. "The Debate on the Standardization of Statistics Maps and Diagrams (1857–1901): Elements for the History of Graphical Language." *Cybergeo: European Journal of Geography* 65. https://doi.org/10.4000/cybergeo.148.

Patriarca, Silvana. 1996. *Numbers and Nationhood. Writing Statistics in Nineteenth Century Italy.* Cambridge: Cambridge University Press. https://doi.org/10.1017/CBO9780511523458.

Penha, Eli Alves. 1993. *A criação do IBGE no contexto da centralização política do Estado Novo.* Rio de Janeiro: IBGE. https://biblioteca.ibge.gov.br/index.php/biblioteca-catalogo?view=detalhes&id=266104. Accessed 19 May 2021.

Ribeiro, Gladys Sabina. 2008. "Nação e cidadania nos jornais cariocas da época da independência: o *Correio do Rio de Janeiro* como estudo de caso." In *Repensando o Brasil do Oitocentos: cidadania, política e liberdade*, edited by José Murilo de Carvalho and Lucia Maria Bastos Pereira das Neves. Rio de Janeiro: Civilização Brasileira.

Sá, Dominichi Miranda de. 2009. "Uma interpretação do Brasil como doença e rotina: a repercussão do relatório médico de Arthur Neiva e Belisário Penna (1917–1935)." *História, Ciências, Saúde- Manguinhos* 16 (Suppl. 1):183–203. https://www.scielo.br/pdf/hcsm/v16s1/09.pdf. Accessed 19 May 2021.

Santos, Julio Cesar Paixão. 2018. "A circulação das ideias de estatística no Segundo Reinado: periodismo, discurso científico e ciências biomédicas no Rio de Janeiro (1840–1870)." PhD dissertation, Fundação Oswaldo Cruz. http://www.ppghcs.coc.fiocruz.br/images/teses/tese_final_JulioPaixoSantos.pdf. Accessed 19 May 2021.

Santos, Luiz Antonio de Castro. 1985. "O pensamento sanitarista na Primeira República: uma ideologia de construção da nacionalidade." *Dados* 28 (2): 93–209. https://drive.google.com/file/d/1jBZHHikOB6fxNvYAu-aVFCsMop5x8qvZ/edit. Accessed 19 May 2021.

Senra, Nelson. 2006. *História das estatísticas brasileiras: Estatísticas desejadas (1822–1889).* Rio de Janeiro: IBGE. https://memoria.ibge.gov.br/publicacoes/historia-das-estatisticas-brasileiras.html. Accessed 28 September 2020.

———. 2007. *História das estatísticas brasileiras: Estatísticas legalizadas (1889–1936)*. Rio de Janeiro: IBGE. https://memoria.ibge.gov.br/publicacoes/historia-das-estatisticas-brasileiras.html. Accessed 28 September 2020.

———. 2008. *História das estatísticas brasileiras: Estatísticas organizadas (1936–1972)*. Rio de Janeiro: IBGE. https://memoria.ibge.gov.br/publicacoes/historia-das-estatisticas-brasileiras.html. Accessed 28 September 2020.

Silva, Joaquim Norberto de Souza e. 1986. *Investigações sobre os recenseamentos da população geral do Império e de cada província de per si tentados deste os tempos coloniais até hoje*. São Paulo: IPE/USP.

Valverde, Orlando. 2006. "Fábio de Macedo Soares Guimarães: *in memoriam*." In *O pensamento de Fábio de Macedo Soares Guimarães: uma seleção de textos*, edited by Nelson de Castro Senra. Rio de Janeiro: IBGE. https://biblioteca.ibge.gov.br/biblioteca-catalogo.html?id=229363&view=detalhes. Accessed 19 May 2021.

CHAPTER 4

Portraits for an Exhibition: The Making of a Statistical Culture for Public Life in Mexico During the Time of the *Dirección General de Estadística*, 1882–1922

Laura Cházaro García

In this chapter, I analyse the practices that led to the production of official statistics in Mexico between the late nineteenth and early twentieth centuries. The analytical approach contemplates two dimensions. First, I examine the bureaucratic practices and knowledge behind the establishment of the *Dirección General de Estadística* (DGE, General Statistics Office), the first institution to produce national statistics, and its transformation in 1922 into the *Departamento de la Estadística Nacional* (DNE, Department of National Statistics) during the post-revolutionary government of Álvaro Obregón. Second, from a local and biographical

L. Cházaro García (✉)
DIE-Cinvestav-IPN, Ciudad de México, Mexico
e-mail: chazaro@cinvestav.mx

C. T. Lanata-Briones et al. (eds.), *Socio-political Histories of Latin American Statistics*, Studies of the Americas,
https://doi.org/10.1007/978-3-030-87714-9_4

perspective, I study the administration of Dr. Antonio Peñafiel (1839–1922), the DGE's Director for 18 years, whose statistical practices embodied the mindset of the bureaucrats who produced numerical data towards the end of the *Porfiriato* (Cosío Villegas 1985; Cházaro García 2000).[1]

Before the DGE's creation in 1882, statistical records in Mexico consisted of fragmentary surveys organised by state governments or produced for political events like elections. The question that arises is why it took so long to begin to produce statistics through a collaborative, nationwide programme capable of overcoming the disregard and illegitimacy with which statistics were viewed by politicians and public opinion alike.

So far, studies on Mexico that analyse national statistics (Lozano-Meza 1991; De la Peña and Wilkie 1994; Mayer 1999; Urquiza 2020) have focused on institutional aspects and on the processes of knowledge production. Statistics, according to these works, were crucial in consolidating twentieth-century nationalisms because they created standardised numerical outputs and arguments that shaped the 'identities' of the subnational territories (Anderson 1993; Patriarca 1996; Porter 2009). These findings are significant, but this work follows a different path. As my interest focuses on the conditions of possibility of national statistics, I rather examine the practices of civil servants and bureaucrats aiming at rescuing the subjective and contingent features of their actions and knowledge. In so doing, I attempt to explore the cultures that shaped the Mexican national statistical system (Desrosières 2010; Prévost and Beaud 2012).

The quantification of social life was not created, or introduced, into people's lives spontaneously, nor did measurements and the things measured appeared uniformly or homogeneously. Broad generalisations regarding political life and science concerning Latin America are hard to make. That is why I pursue two objectives: to problematise the technologies of information collection and to identify the data that those practices created (Von Oertzen 2017; Müller-Willie 2017). To discover how official statistical data was gathered and produced, I explore the grey and

[1] The term *Porfiriato* derives from the name of President Porfirio Díaz, who governed Mexico between 1876 and 1910 through a 'liberal-conservative' coalition with permanent support of the military.

incomplete censuses as well as the statistical archives (Von Oertzen 2017; Agar 2003).

If we accept, quoting Gilles Deleuze and Félix Guattari (1980), that quantification is a means that states employ to impose their authority; then they "striate" smooth and unmeasured spaces, creating metric dimensions, to order people and territories. From the late nineteenth to the early twentieth century, various ways of striating population and the economy were tested. However, those quantifications did not emerge through impersonal or neutral techniques nor as standardised verification metrics (Porter 1995, 94). Throughout this period, I found that in Mexico the production of statistics involved more disagreements than agreements between citizens and bureaucrats, and that the knowledge gathering process was always embedded in multiple negotiations, mainly between surveyors and the surveyed. In fact, the most noteworthy aspect of the production of official statistics was the resistance of both parties. Moreover, respondents resisted to being registered because this meant being placed in categories foreign to their everyday experiences; they rejected being suddenly and unwantedly transformed into members of a family or a professional category like "professor" or "merchant" (Porter 2009, 31). But bureaucrats also resisted. The members of President Porfirio Díaz' governing elite rejected being in charge of the production of statistics arguing that these tasks were not part of their official duties. Following Theodore Porter (1995), those officials were behaving as *gentlemen*.

The life of Dr. Antonio Peñafiel shows that the consolidation of statistics as public knowledge did not emerge by simply institutionalising this discipline or adopting technologies suited to processing large databases. Rather, this process demanded accounting skills and the virtues of the hobby of collecting that Peñafiel and his collaborators mobilised to forge the path that would overcome political resistance to official statistics in Mexico. Trained as a physician, Peñafiel was familiar with medical statistics but, more importantly, had the skills of an antiquarian. He worked at the *Museo Nacional de México* (National Museum of History, later the *Museo de Antropología e Historia*, Museum of Anthropology and History) collecting and curating antiquities and headed the creation of the show "Ancient Mexico" for the *Exposition Universelle* of 1889 in Paris. My aim is to elucidate how that generation of 'gentlemen' bureaucrats mobilised the power of inventory and collection as well as practices of cataloguing and surveying (Von Oertzen 2017).

This generation of DGE civil servants developed a system of affection and hierarchies based on elite relations to produce statistics. This was not a bureaucracy obsessed with mechanical technologies for processing large numbers. Rather, they produced official data taking advantage of the human qualities of individuals who were known more for their honesty and position within the political elite than for their knowledge of sophisticated counting techniques. The DGE produced statistical series on population and the economy, organised them in charts and graphs, and exhibited and disseminated them in publications called *Anuarios* (yearbooks) and *Boletines* (bulletins). These publications were truly graphic technologies, with tables that condensed contemporary knowledge and served as political symbols for governments that aimed to portray images of civilisation and progress. The charts and graphs offered illustrations of the population that translated 'the indigenous' into numbers and standardisations, especially for consumption in Europe. As Ian Hacking (1986) and Alain Desrosières (2010) show, creating official statistics entailed turning numerical classifications or magnitudes from the natural or social world into legitimate things or objects, and in this case, creating practices and images of 'civilisation'. The era of Peñafiel and his social class was one of permanent tension between a world built on charisma and trust and the pulsion to develop routines and techniques to systematise measurings, magnitudes, and figures, standardising the graphs that would portray Mexico as 'modern'.

In the first two sections of this chapter, I describe the political and cultural environment that brought the bureaucrat-gentlemen of the DGE to life, a world characterised by the absence of national statistics. In the third section, and after analysing the place of these statistical gentlemen among the "scientists" and the "wizards of progress" of the *Porfiriato* (Tenorio-Trillo, 1998), I explain how the rarely associated practices of collecting and measuring allowed this elite to create archives and statistical series as well as to develop various paper technologies to show and manage national censuses and population counts. I show how the project of measuring and creating data on national population, territory, and the economy was strongly attached to the political practices that created the story of the Mexican nation: its past, the 'indigenous question', and their (in)civilisation.

Smooth Spaces

After independence, the different Mexican governments showed great interest in organising and collecting data on the population and the territory. The existing statistics included the censuses (*empadronamientos*) carried out in colonial times, like the one organised by Viceroy Conde de Revillagigedo (1793–1796), and the tables and calculations compiled by Alexander Von Humboldt published in his praised *Ensayo sobre la Nueva España* (1811). However, getting to know the population of the new nation implied updating those counts and giving them a national scope. Between independence (1825) and the start of the *Porfiriato* (1876), both liberal and conservative governments devoted great efforts to formulating laws and decrees to organise population censuses. But a myriad of obstacles stood in the way of researching the population of the young republic. Before the *Porfiriato*, censuses and statistics were fragmentary and lacked legitimacy among the general population. Politicians and officials could not agree on how to best apply such measures. This situation has led several historians to describe the discipline and practice of statistics—at least to the mid-nineteenth century—as an accumulation of failures, frustrations, and misunderstandings (De la Peña and Wilkie 1994, 35; Mayer 1999; Arrioja 2016).

Despite the setbacks, the doubts of the public opinion, and the fragmentation of initiatives, diverse actors, amateurs, and scientific associations produced and published 'statistical studies' on the population during the nineteenth century. To begin with, there was an audience that consumed and circulated statistical figures: newspaper readers, professionals—mostly physicians and engineers, usually attached to scientific societies—businessmen, and investors. They needed statistical data for diverse reasons, often to support their research or political interests. Groups that stood out among those consumers/producers were societies like the *Sociedad Mexicana de Geografía y Estadística* (Mexican Society for Geography and Statistics, SMGyE), the *Academia Nacional de Medicina* (National Academy of Medicine, NAM), and the *Consejo Superior de Salubridad* (Superior Health Council, SHC) (Lozano 1991; Azuela 1996; Mayer 1999).

The SMGyE—which grouped specialists in history, geography, and statistics, as well as bureaucrats and politicians—was the most prolific institution in terms of producing and publishing statistical studies. Its members ranked among the forerunners in the field. One example was

José Joaquín Gómez de la Cortina, governor of Mexico City between 1835 and 1836, and author of *Ensayo sobre la población* as well as other works on scientific and statistical topics (Mayer 2003). Another prominent member of the SMGyE was the engineer and geographer Antonio García Cubas, editor of several statistical compendiums like *Noticias Geográficas y Estadísticas de la República Mexicana* published in 1857 and *Apuntes relativos a la población mexicana* released in 1858 (De la Peña and Wilkie 1994, 64). Other active members were the mathematician and astronomer Francisco Díaz Covarrubias, the engineer and mathematician Francisco Jiménez (INEGI 2008, 5–6), and Jesús Hermosa, author of the *Compendio elemental de Geografía y Estadística de la República Mexicana*, published in 1870 (Lozano 1991).

The works of the members of the SMGyE shared certain features. They conceived of statistics as a 'true' description of a country's natural and political wealth. Hermosa considered "statistics the sister of geography" and sought to describe the national territory as a landscape filled with data. Although statistics was not deemed an exact science, it was "one of the most important foundations of politics, political economy, and the enhancement of all peoples" (Hermosa 1870, i–ii, author's translation). These authors described the population using numbers, usually based on data from 'authorities' like Alexander Von Humboldt or fragmentary counts taken by local governments, as in the case of Juan José Martínez de Lejarza's *Analísis estadístico de la provincia de Michoacán* published in 1822. These numbers described the land as part of a narrative, spoke of nature and politics, turned rivers, mountains, lakes, and temperatures into part of the moral history of the peoples.

Physicians organised in the NAM and in the SHC also produced and consumed statistical data. The SHC, also formed by engineers and veterinarians, was an agency of the State Department (*Secretaría de Gobernación*) that gave its members the capacity to intervene in issues related to health, hygiene, and epidemics. From their hospitals, physicians gathered statistical data on disease and mortality that was usually published by the NAM in its *Gaceta Médica de México* or in the SHC's *Boletín*. Albeit circumscribed to the capital city, the data gave those agencies enormous political power within public health (Cházaro García 2018). For instance, the design of the form employed in the 1882 general census of Mexico City was entrusted to the physician Agustín Reyes, former president of the SHC and a city government councilman from 1879 to 1880. These were the same years in which Peñafiel would begin his headship of the DGE that lasted until 1910.

In addition to doctors, engineers, and veterinarians, some private agents also contributed to the emerging market of statistical figures on Mexico's population and economy. Their initiatives used to fill in the statistical gaps that existed at the time, thus attending to the needs of both the state and the federal governments. Basilio Pérez Gallardo, a member of the SMGyE, is an illustrative case (Medeles 2011). He produced several statistical studies, one at the request of President Benito Juárez that was published as *Cuadro estadístico de la división territorial de la República Mexicana* (1873). He was aware of how important statistics was "for all branches of administration" and that much of the demographic and fiscal information available in Mexico were mere "approximations" or "improvisations" (Pérez 1873). That is why he insisted that the government of Mexico City had to obtain data directly from the actual sources. Later, accompanied by his son, Pérez went street-by-street to compile information for composing his *Plano topográfico descriptivo de la Ciudad de México* released in 1878, which accounted for the number of inhabitants, judges, and municipal agents by neighbourhood (*barrio*) or ward (*cuartel*). That study was sold to the government for 1600 pesos.[2]

Another case was that of Emilio Castillo Negrete who, in 1880, sold to the government of the country's capital 50 copies of his *Cuadro sinóptico de la producción agrícola de la República Mexicana* for 125 pesos. This text was a census of agricultural production of rural ranches and haciendas.[3] In 1886, when the DGE already existed, the Ministry of Development (*Secretaría de Fomento*) contracted the services of the Mercantile Agency of the Mexican Republic (*Agencia Mercantil de la República Mexicana*) to obtain data and produce reports on demography and on the economy. The mercantile agency was headed by Alfonso Lancaster, a lawyer for the *Prida Navarro* company, and by José Yves Limantour, who served as Minister of the Treasury between 1892 and

[2] Archivo Histórico de la Ciudad de México (AHCM). "Pérez Gallardo Basilio propone un cuadro topográfico descriptivo de la Ciudad de México", 28 January 1878, f. 1. Sección Estadística, Vol. 1031, Exp. 39. Fondo Ayuntamiento, Gobierno del Distrito Federal (Fondo AGDF).

[3] AHCM. "Se consulta el gasto de 125 pesos para la compra de 50 ejemplares del Cuadro Sinóptico de la producción agrícola de la República Mexicana", 1880, 9 fs. Sección Estadística, Vol. 1031, Exp. 41. Fondo AGDF.

1911 and was the living incarnation of the gentleman bureaucrat of the *Porfiriato*. The agency sold statistical series prepared with data collected by public bureaus, like the Ministry of Development and the Treasury, as well as by local governments, such as the *ayuntamiento* of Mexico City.[4]

Following Deleuze and Guattari's (1980) notion of smooth and striated spaces, the statistical enumerations compiled by these scientific associations, professionals, and other agents offered a "smooth" classificatory order. The series were numerical lists of population, concerning births, deaths, and marriages, most of the time fragmentary and retrieved from different sources, like parishes, surveys carried out in the state or municipalities, and the Civil Registry (Arrioja 2016; De la Peña and Wilkie 1994).[5]

Between 1853—when the Ministry of Development was created—and 1882—when the DGE was created—government representatives relied on third parties to collect, organise, and systematise statistical data. The efforts of scientific associations and independent agents reveal that the state bureaucracy did not exercise a monopoly on the production of statistics. These circumstances force us to think of the epistemological and political complexities involved in producing such information (Müller-Willie 2017, 112), as well as to ask about the genealogy of the statistical technologies that state officials ended up embracing, in particular what political order, and order of knowledge, they were reproducing. According to Jon Agar (2003, 57), the significant moment when nineteenth-century citizens encountered the state was when they filled out an official form. Through these actions, statistics began to acquire meaning and practical weight. But such description contrasts with what happened in Mexico, where data was systematised outside official institutions, although those outsiders depended on the information collected and provided by government agencies.

This conundrum can be clarified by introducing the category of "unexpected citizens", developed by Adriana Acevedo and Paula López (2012). When speaking of 'citizens', we usually think in abstract terms, of people who participate in a linguistic and political community, subjected to a particular jurisdiction (Anderson 1993). For nineteenth-century Mexico,

[4] AHCM. Secretaría del Ayuntamiento, Sección 4ta, no. 2, 1886–1887. Sección Estadística. Vol. 1031, Exp. 46. Fondo AGDF.

[5] AHCM. Luis C. Barreda, Oficial Mayor, Secretaría. Supremo gobierno, Mexico, 3 July 1838. Leg. 1, Censo, f. 1 and 1R. Fondo AGDF.

Acevedo and López (2012, 22) argue that analysing the question of citizenship means renouncing to models and aims of universality to focus on local experiences, though those realities hardly correspond to juridical definitions or to the models used by political science. These "unexpected citizens" are the outcome of local experiences, particular political cultures, and extra-legal practices. In Mexico, the production of national statistics involved, precisely, 'unexpected' citizens, since neither the officials nor the people were related through mandates of measuring and being measured. Thus, we can also speak of 'unexpected statistics'. Though governments envisioned the possibility of transforming people into a governable resource that could be directed and educated (Patriarca 1996, 7), few government officials were convinced that this could be achieved through the language of measurements and statistics. The elite and the bureaucracy that should have carried out this work had a political and an intellectual mindset alien to the drive to survey the population and generate statistical series based on forms and questionnaires. They did not believe in the possibility of converting what they governed—the population and the economy—into abstract numbers. Though all people were legally equal, not everyone participated in the political community in equal terms: women, indigenous peoples, and the poor could not be considered citizens, and consequently, measured or quantifiable in those terms (Hale 1991; Palti 2005).

Hence, the sources that gave legitimacy to the statistical data emerged through tense negotiations between those gentlemen bureaucrats and 'the people'. The bureaucrats—educated in natural and national history—were acquainted with collection techniques used by collectors and museums. As we will see, such knowledge and its practices led them to develop a logic for ordering data in charts, usually produced for public exhibitions. The population, in turn, came gradually into terms with this drive, allowing officials to get into their domestic life, providing data, and letting themselves to be measured.

In the following section, I examine the conditions that made it possible to striate a population and a territory that—until then—had been managed as a smooth landscape with abundant resources and where the idea of magnitude was practically absent. The government's goal was to create a quantified ordering of the population, an ordering made up by magnitudes and graphs that would express the content of the nation. By gathering statistics that would striate the territory, bureaucrats sought to create conditions for administering, intervening in, and controlling the nation's affairs. But reaching this goal required negotiating with people and administrators on how to handle the numbers and images produced, especially regarding those identified as indigenous. What follows is an exploration of the vicissitudes that emerged in creating that order and translating it into acts of government (Desrosières 2010).

The Striated and the Resistance of the Gentlemen

From the 1850s onward, conservative and liberal governments passed a number body of laws that allowed to know and display in figures and maps the sources of production of goods and the size of the population. However, that information was produced in fragments and was usually compiled by scientific associations, investment groups, businessmen, and private actors, none of whom were interested in crafting portraits that accounted for the nation as a whole. The Ministry of Development was created in 1853 during Santa Anna's government. A year later, the government established a law that fined citizens who refused to participate in censuses. In 1857, the official recording of life events in the Civil Registry became compulsory, as judges were made responsible for registering births, deaths, and marriages. To apply these norms, the Ministry of Development found itself forced to form a specialised bureaucracy, a corps of officials and technicians responsible for generating data for public administration.

After Santa Anna's defeat during the Revolution of Ayutla in 1854, Benito Juárez and the liberals came into power. Between 1857 and 1867, that group deployed the so-called *Reforma*, which opened a critical and sometimes violent political process that ended with a deep social revolution. The *Reforma* revealed the need for an expert bureaucracy capable of fomenting and administering the wealth of the republic, and rationalising the state administration over the land and the population. According to María Cecilia Zuleta (2000, 5), the Ministry of Development sought to move away from a *consultative* administration, which was merely a consumer of data, towards one that would *actively* link the financial possibilities of the federal treasury by expanding public spending (Cosío Villegas 1985). In 1853, the Ministry opened a Statistical Section (*Sección de Estadística*), re-enacted in 1876, but its functioning was still based on *consultative* rationality. The section was limited to compiling information produced by administrative units, whether agricultural, industrial, educational, or judicial (Blanco and Omar 2011). Because public administration officials resisted to gathering data themselves, the distance between their agencies and the objects that were to be measured widen. As other historians demonstrate, European nation states boosted the production of official statistics through specialised bureaus (Patriarca 1996; Desrosières 2010). In Mexico, several members of the SMGyE demanded the creation of a bureau of statistics that would rectify the fragmentary efforts to elaborate statistics and maps (Mayer 2003). However, the establishment of the DGE did not automatically begin the regular production of official statistics.

Creating statistical facts or data requires multiple intellectual and material acts; in other words, statistical facts are manufactured (Agar 2003; Poovey 1998). In the Mexican case, the archives of the liberal administrations in office between 1850 and 1876 reveal that the official bureaucrats, due to diverse reasons, resisted taking to the streets to survey citizens. The analysis of their practices shows that they were not trained to produce statistics, which probably reinforced the perception that these tasks were beyond their areas of expertise. More importantly, they felt that these activities were unworthy of their professional standing. The political training of these gentlemen bureaucrats hindered them from generating data through large-scale questionnaires. As a result, their professional performance depended on personal relations that were usually contingent and unsystematic; their knowledge of how to measure the population and the territory was imbricated on those relations. In their worldview, the population, the economy, and politics were not *objects* with measurable features (Porter 1994, 391; Poovey 1998; Desrosières 2010).

This becomes clear when analysing the actions of civil servants at different levels of the bureaucratic chain—federal, state, and municipal—who participated in diverse censuses taken in Mexico City. Examples include the *Censo de personal y de establecimientos mercantiles e industriales* (Census of Mercantile and Industrial Personnel and Establishments) planned by the State Department in 1854,[6] as well as the 1868 *Padrón Municipal* (Municipal Census), ordered by the city government and carried out by civil registry judges, and the 1882 census, requested by the Statistical Section of the city council of Mexico City. In all these operations, the residents of Mexico City were surveyed using different forms and techniques. The 1854 census was carried out by the then governor

[6] "Bando para el empadronamiento general de la Ciudad de México", Antonio Diez Bonilla, 1854, AHCM, Leg. 2, 1854, f. 1 and 2.

of the State of Mexico Antonio Diez Bonilla, while the other two were ordered by the governor of the state and by the city council of Mexico City.

A review of the processes followed in those censuses shows a scarce participation of well-trained census-takers, because the candidates had to demonstrate that they were honourable people (*gente proba*) 'with recognised authority' to guarantee the obedience of the population. They also had to be able to read maps and know the city's streets and neighbourhoods. Only a handful civil servants satisfied those criteria, and those that did, like judges and government employees, resisted to carry out those tasks, showing their unwillingness to participate in producing the required statistical data.

The classic argument that emphasises citizens' resistance to surveys tends to hide the other side of the equation, the reluctance showed by census-takers. In reaction to the 1854 enquiry, city councilmen (*regidores*) stated their refusal to "carry out the work related to the creation of the register of inhabitants". In their testimonies, they claimed that census-taking impeded them from attending to their businesses and day-to-day jobs. Their attitude led the governor of Mexico City to petition the State Department to "immediately" hire replacements, people who would be "as worthy as those whom they would replace".[7] In the case of the 1868 Municipal Census, the government of Mexico City requested civil registry judges to collect data on the population of the areas of their jurisdiction. The judge in charge of the Second Tribunal rejected the tasks, arguing that the government was unaware of the "difficulties" it was facing: lack of properly trained personnel and the authorities' failure to resolve such basic questions as "the forms to be used, the number of people to employ, and the resources for paying them".[8] Between 1881 and 1882, Agustín Reyes—a Mexico City councilman and author of the *Proyecto de censo de la Ciudad de México de 1882*—expanded upon the judge's objections, writing that the authorities lacked "precision in naming the people commissioned to conduct the census (…)", and failed to offer "protection and support". Reyes emphasised that census-taking was not a "simple"

[7] AHCM, Leg. 2, 31 July 1854, f. 4 (author's translation).

[8] Juzgado 2°, carta a la Secretaría del Ayuntamiento constitucional de México, 27 November 1868, f. 7. AHCM, Leg. 5, f. 7; "Sobre que se remitan al C. Superior Gobierno los últimos datos relativos al padrón de la población que ecsiste en la Secretaría", AHCM, Leg. 1, f. 1.

job, or a task "easy to perform".[9] Census-takers, he continued, did not know how to "negotiate" with people and were unfamiliar with the areas where the poorest resided, spaces that were growing vertiginously in the late nineteenth century (Rodríguez-Kuri 1996, 190). Moreover, for Reyes

> ... experience suggests that [success] requires civic responsibility and knowledge of the area in order to know how to suitably treat each class of the society so as to obtain with accuracy the number of inhabitants residing in each house as well as the other aspects and data they consider necessary; without [this experience], as on other occasions, some areas are not registered, due either to lack of knowledge of the place, or the difficulty in reaching certain neighbourhoods and suburbs when carrying out said task.[10]

In Reyes' view, census-taking should not be considered a part-time or supplementary job, neither a *distinction* nor a *voluntary* activity.[11]

Census-takers were involved in diverse disputes well into the twentieth century. Under Peñafiel's direction, the DGE was an organisation that alternated between having volunteers from "the best social positions" and having paid staff. In 1895, when the DGE organised its first national census, the Minister of Development, Fernández Leal, decided that they would rely on the work of non-paid staff as this would ensure that the DGE obtains "trustworthy results". His goal was to employ people who

> ... exercise authority [where] residents are used to respecting and obeying them (...) There is no reason, as it is usual for common censuses, to think of naming poor people who receive a small gratification but perform their work with [minimal] consciousness of what they are doing. In the 1890 [Mexico City] census, census-takers were people of the best social position, for this reason their results were trustworthy.[12]

The DGE took surveys relying on both the elite and 'poor' people, but the latter carried out the fieldwork, regardless of their knowledge of statistics. The results, however, acquired authority and credibility given

[9] A. Reyes, "Proyecto para levantar el censo de la Ciudad de México", September, 1880, AHCM, Leg. 5, f. 8r.

[10] A. Reyes, AHCM, Leg. 5, f. 2 and f. 2r, author's translation.

[11] A. Reyes, AHCM, Leg. 5, f. 8r.

[12] AHCM, Caja 65, Exp. 20, f. 3, author's translation.

the knowledge of their superiors. In this sense, the statistics generated by the DGE were the result of negotiations among civil servants, the humble workers, and the population. The best-trained bureaucracy was devoted to systematising data and to publishing the results of censuses and statistical series on the country's economy. The DGE's publications, which were widely distributed and exhibited, translated the smooth landscapes of the territory into magnitudes and figures. Bureaucrats privileged two ways of doing statistics. For census-taking, they used forms to survey families. To account for economic activities, they requested information annually from different ministries and, more generally, state governments. Both strategies were designed to create national statistics and to guide the administration in planning the nation's economy. Although these strategies were intended to generate reliable data that guaranteed impartiality, the information was still based on practices in which officials acted through personal relations and judgements. Instead of following rational norms and methods, there was a strong sociopolitical element behind the elaboration of data (Porter 1995, 102).

Practices of Collecting Statistical Data During Peñafiel's Life

General Carlos Pacheco Villalobos (1839–1891) was the Minister of Development who promoted the creation of the DGE during the presidency of Manuel González (1880–1884). As a man of his time, Pacheco was concerned with producing "official" or "administrative statistics", a discipline that required combining "science and the art of government". He defined this as a science "inclined towards practice". Those who made statistics, he wrote, must focus on "conclusions or deductions that can influence the political government of a nation" for the "good of the people" (Pacheco 1882, 126, author's translation).

To organise the DGE, Pacheco created an auxiliary commission with two members: Antonio Peñafiel and Francisco Ramírez Rojas. Ramírez was a civil servant of the Ministry of Development, enrolled in the Section of Industry, Commerce, the Mint Office, Telegraphs, and Weights and Measures. He was assigned the task of pondering "the administrative organisation and service" of the new office. Peñafiel was instructed to define the matters that would be included within "national statistics" (Pacheco 1882, 128). Before his appointment, Peñafiel had worked as a physician. He was an active member of the NAM and a founding member of the *Sociedad Mexicana de Historia Natural* (Mexican Natural

History Society) and of the National Museum (INEGI 2010, 20). It is not clear why he was chosen for the DGE since he had no experience with statistics apart from publishing “Ensayo de análisis estadístico sobre lesiones” (Essay on a Statistical Analisis of Injuries) in the *Gaceta Médica de México* (Peñafiel 1883). His research interests and publications were concerned mainly with natural and ancient history, the hobby of collecting, archaeology, and indigenous languages. In 1868, while he was still quite young, Peñafiel was hired by the National Museum as a tutor of zoology and botany. He collaborated with naturalists and zoologists of the time, such as Antonio Isaac Álvarez, Gumensindo Mendoza, and Antonio del Castillo, with whom he would later participate in establishing the *Sociedad Mexicana de Historia Natural* (Cuevas Cardona 2006; Rodríguez de Romo et al. 2008, 359).

Peñafiel was closely related to the elite of civil servants and intellectuals of the *Porfiriato* and understood that he could capitalise on those relations to carry out his projects. Ramírez died just a year after beginning organising the DGE, leaving Peñafiel as the only director. As the leading figure in the field of national statistics, Peñafiel published several works through the Ministry of Development, such as *Monumentos del arte mexicano antiguo: Ornamentación, mitología, tributos y monumentos* (1890a, b) and *Nomenclatura geográfica de México*, and *Etimologías de los nombres correspondientes a los principales idiomas que se hablan en la República* (1897). He also edited and reprinted the *Códice Fernández Leal* (1895), *Colección de Documentos para la historia mexicana* (1897–1903), and *Diccionario de las lenguas Tarasca y Michoacana de Fr. Maturino Gilberti* (1901), among others. In those years, a collaboration with Lamberto Asaín, then Deputy Secretary of Development, led to the publication in 1884 of *Memoria sobre las aguas potables de la capital de México* (Peñafiel 1884) and *El Lago de Texcoco* (1895), two key studies on the works of desiccation within the Valley of Mexico.

Peñafiel’s involvement in marketing and exhibiting Mexican antiquities included travelling to Europe and to the United States. He took advantage of his participation in those trips to distribute the DGE’s publications. For Peñafiel, the universal exhibitions held in the late nineteenth century represented a showcase for displaying the products of Mexican ‘science’ and history. His attendance to these events aimed to show that Mexicans knew how to govern through scientific politics. As other historians point out (Tenorio-Trillo 1998, 174; Garrigan 2012, 153; Sánchez-Oeconomo 2018), the DGE participated in several universal exhibitions, displaying its *Boletines* and *Anuarios* of national statistics. The Ministry of Development's policy aspired to show, in Mexico and abroad, a quantified image of the country's territory, wealth, and civilisation.

Peñafiel exhibited his works on various occasions, especially at the 1889 Paris Exposition and the 1903 International Expo in St. Louis, Missouri. Perhaps his most outstanding piece was the *Pabellón Azteca* displayed in Paris, a building he designed with the engineer Antonio M. Anza (Peñafiel 1889a, b). Peñafiel's aim with the *Pabellón Azteca* was to display the grandeur of Mexico's pre-Hispanic indigenous peoples, not of his contemporaries. A year later, at the 1900 *Exposition Universelle de Paris*, he took advantage of the occasion to exhibit results of the 1895 population census, winning a gold medal for the DGE (De Mier 1900, 38–40, 43; Sánchez-Oeconomo 2018, 115–116). Double-entry statistics, charts, and tables with census and fiscal data published by the DGE became objects worthy of being displayed, allowing Mexico to introduce itself as a quantified order, like any other European country.

At the end of his life, Peñafiel combined two abilities rarely cultivated together: that of a historian-collector who offered a certain interpretation of Mexico's ancient past with that of an organiser of national statistics. His gaze as a collector of antiquities refined the techniques of gathering, classifying, and ordering statistics. Peñafiel transformed statistics and antiquities into sources that allowed the state to speak in the name of the nation (Deleuze and Guattari 1980). Under Peñafiel's headship, the DGE transformed statistical data into organised charts and hordes of numbers into graphs, adopting the most modern language to offer a global view of Mexico's economy and population. Thanks to Peñafiel's activities as a collector, those charts were integrated into a patrimony that was exhibited inside and outside the country. Though no plan existed, in the late nineteenth century the DGE presented a vision of Mexico derived from technologies unrelated to the sophisticated handling of large numbers.

Administrative Technologies: Surveys, the Hobby of Collecting, and the Publication of Charts

Seeking to determine the best model for national statistics, the Minister of Development asked Peñafiel to study the practices of producing numbers followed in European countries. Hence, Peñafiel examined the models used by Moreau de Jonnés and Ernst Engel who headed the central bureaus in France and Prussia, respectively (Fig. 4.1). He also consulted those used in the United States, but was finally persuaded by the French and German approaches.

This led him to define the DGE's official statistics as follows:

> I. The national census; II. The registry (*catastro*) of urban, rustic, [and] mining property as indicators of national wealth; III. Detailed registry of the nation's agricultural production; IV. Industries of diverse classifications (raw materials and their consumption); V. Trade regarding exports, imports, and between countries; VI. Table of public schools and charitable establishments; VII. Local roads, roadways, canals, telegraphs, and railways; VIII. The course of civil and criminal justice, religions; IX. The contributions of public rent; X. The state of the armed forces and of the navy; XI. All what statistics encompasses in its most important ramifications. (Peñafiel 1882, 132; Medeles 2016, 156–157, author's translation)

To deal with such an enormous research plan, the DGE asked for resources to have a director, three civil servants, four scribes, and two

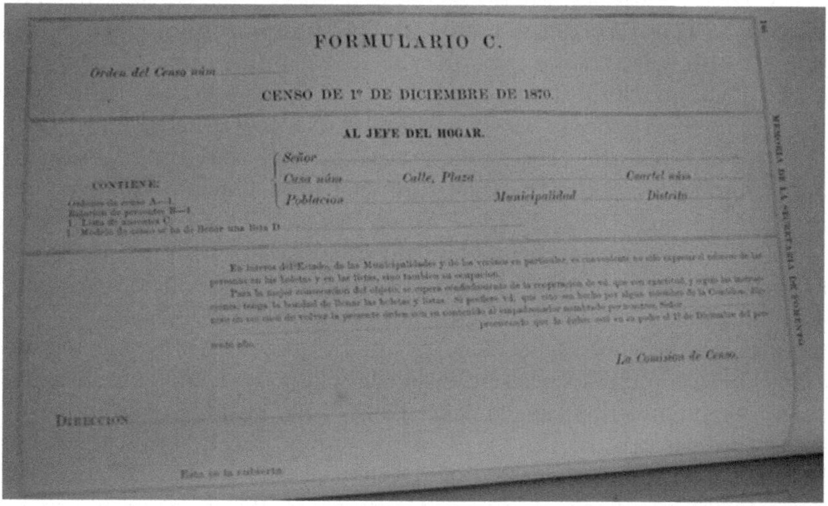

FORMULARIO C.

Orden del Censo núm.

CENSO DE 1° DE DICIEMBRE DE 1870.

AL JEFE DEL HOGAR.

CONTIENE:

Señor

Casa núm. *Calle, Plaza* *Cuartel núm.*

Poblacion *Municipalidad* *Distrito*

La Comision de Censo.

DIRECCION

Fig. 4.1 Example of a German census form (Berlin 1870) that Peñafiel used to create forms for the 1882 Mexico census (*Source* Peñafiel [1882, 146] Courtesy Archivo General de la Nación, Mexico)

clerks. One may ask how it was going to be possible to gather and organise inquiries on all the topics listed in the national statistics plan with such a modest number of staff, or how the DGE was going to be able to collect, organise, classify, and store all that information, and then design charts to display it.

The DGE's publications shed some light on the methods adopted under Peñafiel's headship (Peñafiel 1882, 144; Liceaga 1892, 15–18). In the 1890s, the DGE introduced a system of auxiliary boards, similar to the organisation that the SMGyE had adopted in 1851 (Lozano 1991, 162). The boards were made up of bureaucrats from all levels of government administration: ministries as well as community, municipal, and state officials. The goal was to accelerate the production of data in their territories and jurisdictions. The DGE officials were responsible for coordinating the boards set up across the country and for concentrating, ordering, and disseminating the information produced. In the case of the DGE's personnel, Peñafiel expected to "form auxiliary boards in each state", headed by an agent of the Commission on Statistics with people "well-accepted in the society where they lived" and "recommended by state governments" (Peñafiel 1882, 148, author's translation).

The 1890 census of Mexico City—organised by the DGE and the SHC (Liceaga 1892)—served as a model for the future censuses taken by the DGE. It also acted as the basis for the first national census, conducted by the DGE in 1895. Those experiences were reproduced, with slight changes, in the 1900 and 1910 censuses (DGE 1901; Echegaray 1912). In all these cases, the census-takers were formally hired. They reached the city house-by-house to make sure that even illiterate families were surveyed (Liceaga 1892, 27) and strove to reach even the most remote dwellings. They were supervised by DGE civil servants with the support of the local officials that compounded the auxiliary boards, including police officers, soldiers, and some dignitaries, including the President of the Republic. As usual, census-takers had to be citizens with good reputation in the communities where they lived, enlightened, and committed to the enforcement of the law (Peñafiel 1882, 143, author's translation).

In a regular census, the 'head of the family' received and filled out the forms, returning them to the DGE civil servant by the next day. Those forms contained questions pertaining to the community, to the family, and to each individual. The information requested included the names of the members of the family, their "sex, age, relation to the head of the household, marital status, occupation, religion, language spoken, origins, birthplace, nationality, domicile (if the surveyed was not member of the household), [and] defects [*sic*] such as blindness, deaf-muteness, idiotism, mental disorders" (Peñafiel 1882, 145; Liceaga 1892, author's translations). The DGE usually distributed one printed form per family, not per individual. Peñafiel preferred this way because he considered that in the long run it would save time, since civil servants were not accustomed to handling surveys, and because it would allow obtaining data on collective units, which he deemed equivalent to the family. He claimed that "for Mexico, one form would be enough to gather the characteristics of the family as a *collective* and the isolated individual, in light of the difficulties that would arise from using multiple forms in the hands of authorities unaccustomed to statistical efforts" (Peñafiel 1890a, b, 148, author's translation, italics in the original).

The use of one form per family was also conceived to avoid the problems imposed by the alleged ignorance and disobedience of Mexican people:

> In Mexico, many difficulties must be overcome: the diversity of languages and dialects of the races [...] which reaches 108 (types). The individual form, [Ernst] Engel's statistical ideal, cannot pass the lintel of the huts [inhabited] by indigenous who cannot write; but it may open the way to later making this step easier. (Peñafiel 1890a, b, 141, author's translation)

These censuses were deeply striated exercises that created demographic dimensions and magnitudes never handled before by government agencies. The data depicted a territory that had always been perceived as a smooth, linear space, which was suddenly being ordered in magnitudes and stable categories that imposed the need to update what was known about each locality. So, the DGE's census-takers and auxiliary boards took up the task of "rectifying the number of municipalities, towns, haciendas, and ranches in each state". Using maps, statistic officials marked the point where each locality existed (longitude and latitude), their names (old and new), and added the climatic conditions in each region (Peñafiel 1882, 143; Peñafiel 1884; DGE 1898, iii). Processing all that data required several years of work. In 1884, Peñafiel acknowledged that the public administration "did not know exactly the number of municipal divisions of the towns and other places inhabited over the breadth of the territory of the Republic", key knowledge "for the operations of statistical science" (Peñafiel 1884, i, author's translation).

It is important to note, moreover, that the operations performed during population censuses differed from those used to measure the nation's wealth, like agricultural production, infrastructure, property, mining, and commerce. In this regard, I paraphrase the question put forward by Otto Peust, an official of the Ministry of Development, who asked: how are we going to ascertain, statistically, our political-economic problems? (Peust 1910, 5). The DGE did not produce data on economic matters, it only *collected* it. It is worth noting that there never were forms to survey or gather information on property, industry, mining, or agriculture, since up to 1910, all those reports were provided to other government ministries and agencies. To compile the 1897 *Anuario de estadísticas nacionales*, for instance, the DGE requested information from state governments. In the case of Mexico City—a place that faced the greatest difficulty in generating statistical data—the DGE requested in 1896 information on telegraphs, telephones, railroads, and public finances, as well as on the slaughtering and supplies of cattle, sheep, and pigs. Another aspect that stands out in these and other cases is that the DGE did not have standardised protocols for compiling economic information. Correspondence among civil servants reveals innumerable confusions and misunderstandings regarding what the DGE asked for or wished to know. In some cases, civil servants attached hand-drawn frameworks or *esqueletos* (Fig. 4.2) to their replies to give correspondents an idea of the data to be included in each request.[13]

In the end, favouring this approach towards data collection meant that the DGE did not develop mechanised accounting routines for the country's economic wealth, limiting its role to adding bits and pieces of information produced by other institutions or agencies, like the Ministry of Development, state governments, scientific communities, and professionals. In the *Anuario estadístico de 1900* published in 1901, for example, the chart "Geographic Coordinates of some of the Principal Points of the Mexican Republic" (Fig. 4.3) identified the following 'authorities' as data providers: Francisco Jiménez of the *Comisión Geográfico Exploradora* (Geographic Exploration Commission), Salazar Ilarregui of the *Observatorio Nacional* (National Observatory), and the SMGyE.

At first sight, those charts simply compiled data from multiple sources and authors with no more certification than that conferred by the

[13] AHCM, Vol. 1032, Exp. 71 or 76, f. 2.

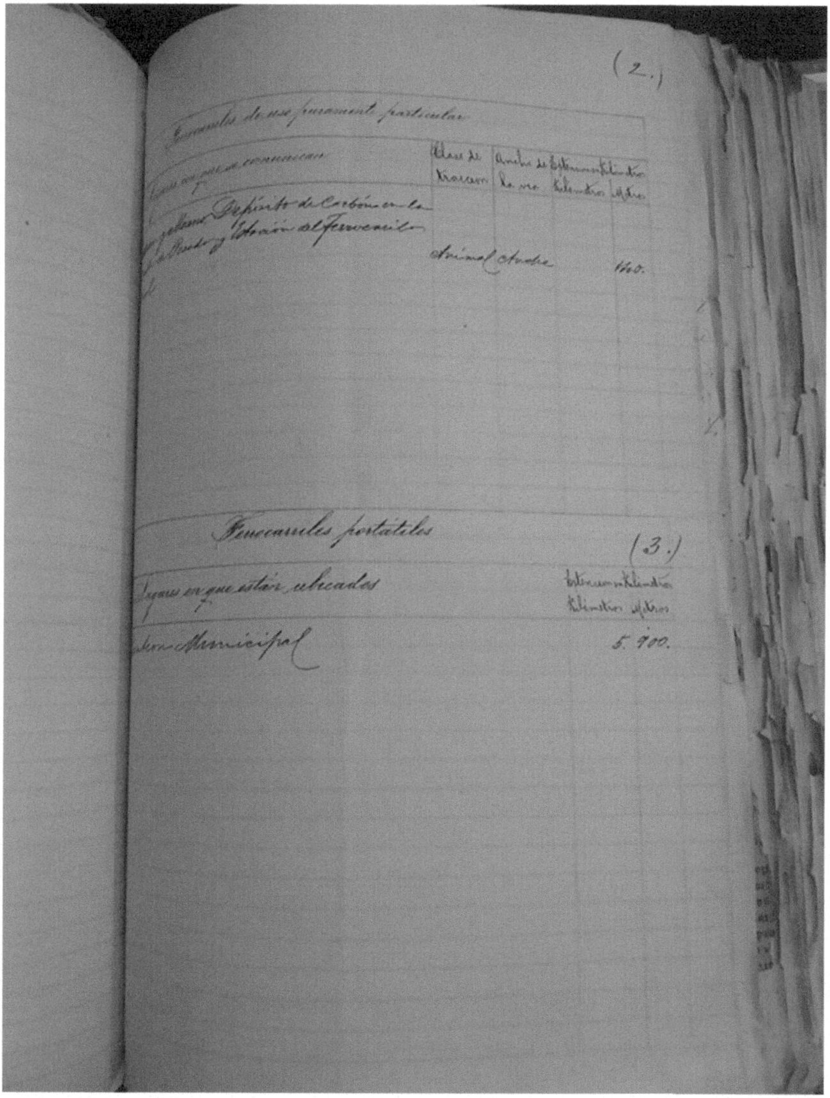

Fig. 4.2 *Esqueleto* (framework) for the civil servants of the Municipality of Mexico City to fill in with the data requested by the DGE in 1896 (*Source* AHCM, Vol. 1032, Exp. 76, f. 2. Courtesy of Archivo Histórico de la Ciudad de México)

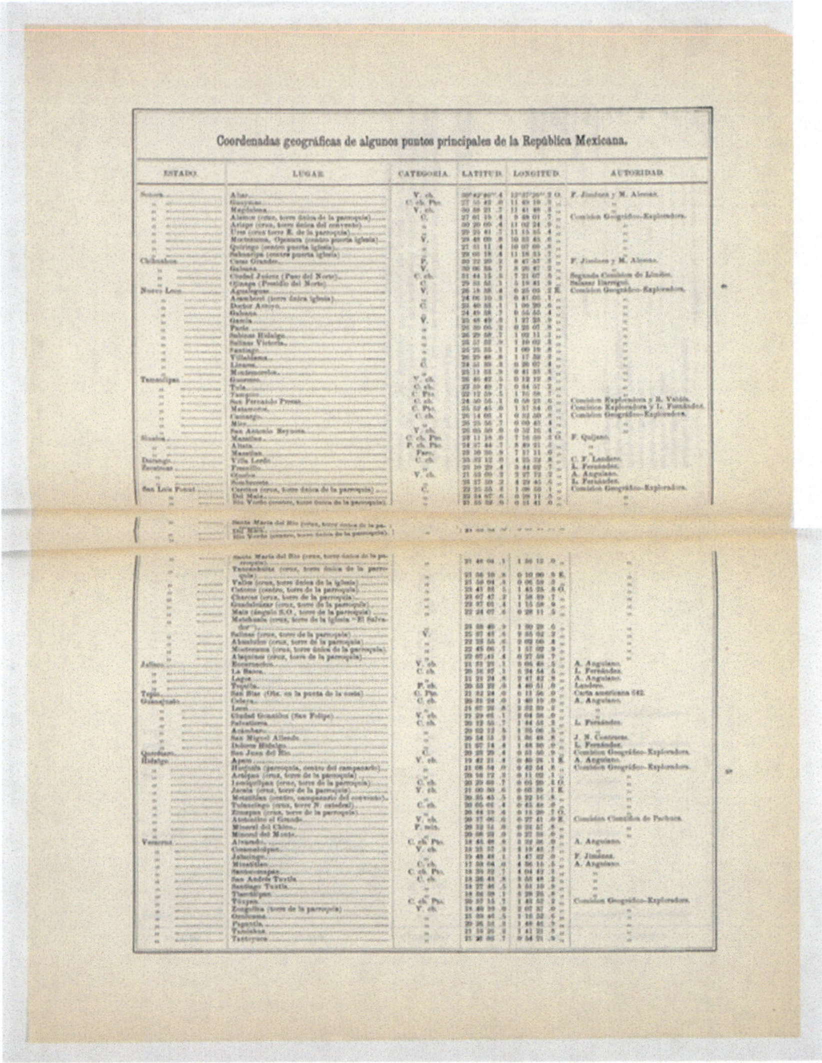
Coordenadas geográficas de algunos puntos principales de la República Mexicana.

ESTADO	LUGAR	CATEGORÍA	LATITUD	LONGITUD	AUTORIDAD

Fig. 4.3 Chart of geographic coordinates prepared for the *Anuario estadístico de 1900* with information provided by engineers and astronomers (*Source* DGE [1901, 4–5]. Courtesy of Bibliothèque Nationale de France)

producer's reputation. Despite its flaws, the DGE generated charts that classified and ordered the data collected and, more importantly, published and exhibited them for national and international consumption. Indeed, the DGE's publications produced tables and graphs of quite extraordinary quality, helping to give quantitative visibility to the economic and social policies of Porfirio Díaz' government. Among those works, we can mention the *Boletín semestral de la República Mexicana*, published between 1888 and 1892, which became the *Anuario estadístico de la República Mexicana* in 1894. Good examples can be seen in the graphs of Figs. 4.4 and 4.5 from the *Anuario* of 1900, a publication that lasted until 1912. The aim behind the design of graphs was to summarise the enormous amounts of data collected, give them coherence, and eliminate any ambiguities that might arise from their diverse origins or production (Tufte 2007, 20).

The DGE's publications displayed graphical technologies, bi-dimensional charts, and maps that organised and gave meaning to data. At a glance, observers could 'read' images or charts that synthesised (with high aesthetic standards) several layers of complex information, making the politics of the government accessible to the general public and open to public debate, particularly between supporters and opponents of Porfirio Díaz's regime. One good example is *Los Estados Unidos Mexicanos, sus condiciones naturales y sus elementos de prosperidad* published in 1893 by Rafael de Zayas Enríquez, a politician from the state of Veracruz. Based on the DGE's statistical inquiries and data, the book was shown at the 1889 and 1900 Universal Exhibitions in Paris as an iconic reflection of the success of the economic performance of the *Porfiriato*. Likewise, the *Anuarios* and the *Cuadro Sinóptico de la República Mexicana,* both published in 1910, illustrated the economic and demographic variables shown in graphs and maps. This way the territory and population were striated. They were transformed into magnitudes, graphs, and other visual objects by means of technologies capable of translating 'the political' into a quantitative order.

The data-gathering techniques developed by the DGE, however, ran into certain problems. Like it happens with collectors, the agency functioned with a permanent state of incompleteness because its research relied on civil servants who failed to provide information or to provide it on time. This was one of the reasons why the *Boletines semestrales de estadísticas* ceased to be published biannually to become, from 1894, yearbooks. But the problems persisted. "Sadly, most of the charts that

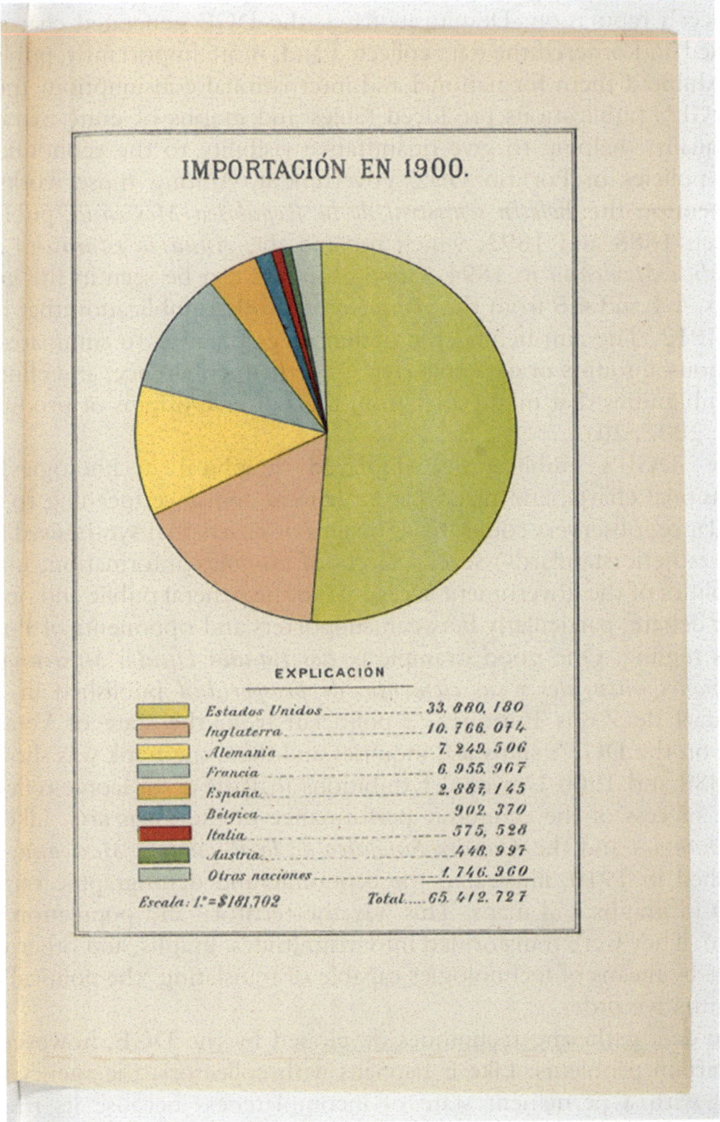

Fig. 4.4 Visual display of statistics on imports in the *Anuario estadístico de 1900* (*Source* DGE [1901, 161]. Courtesy of Bibliothèque Nationale de France)

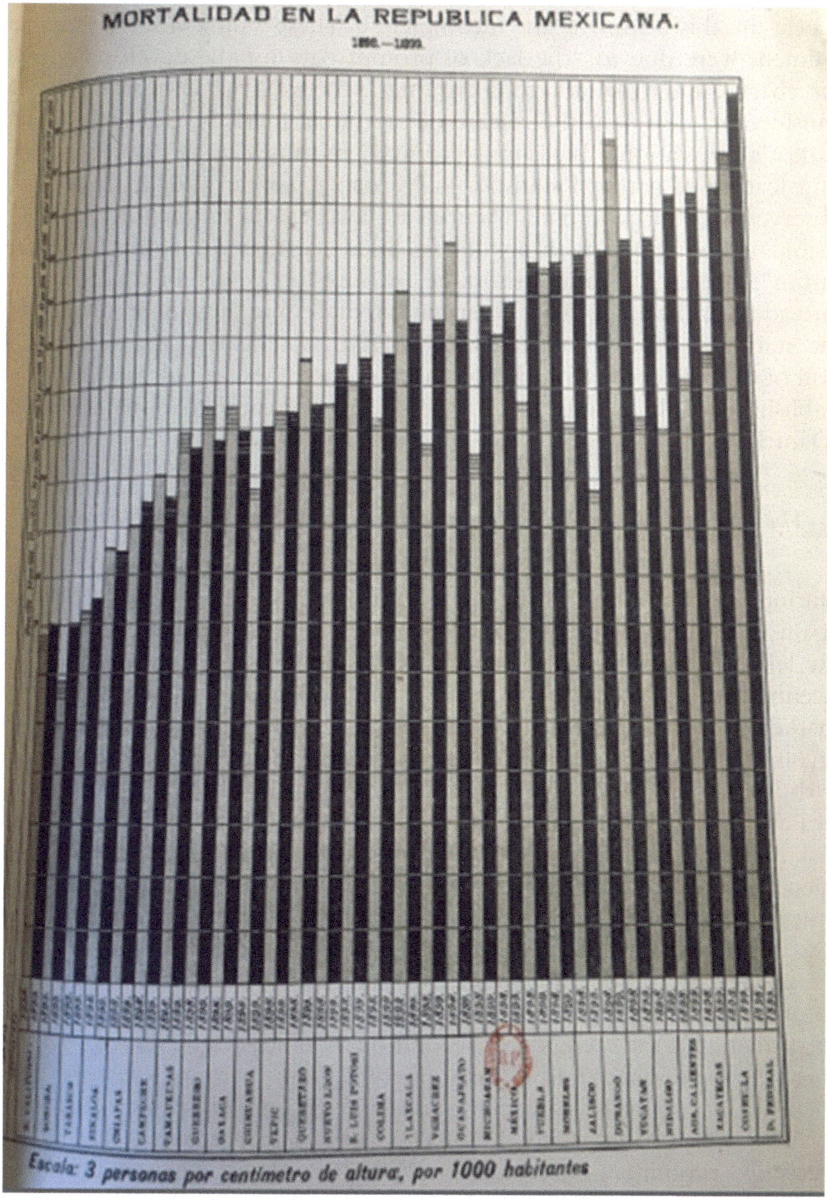

Fig. 4.5 Mortality rates in the *Anuario estadístico de 1900* (*Source* DGE [1901, 171]. Courtesy of Bibliothèque Nationale de France)

appear in this *Anuario* are incomplete". These "difficulties", Peñafiel claimed, were due to "the lack of prompt data for the development of the charts" and "the laborious printing [process]", which resulted in "a considerable delay" in the release of the publications (DGE 1894, i, vi, author's translation). Inaccuracies due to missing data became a recurring feature of national statistics in Mexico. Garrigan (2012, 169–170) observes that the government tended to compensate the gaps and imprecisions in its reports with a political discourse that promised "a better future", one in which any shortage or omission would be transformed into additional wealth. What is certain is that despite their incompleteness, the statistical works produced by the DGE systematised the accumulation of experiences in data creation and technologies for representing and publishing graphs, reinforcing the possibility of creating national statistics (Daniel, 2014).

The Strength of the Impersonal and the Organisation of National Statistics

During the celebration of the centenary of Mexico's independence from Spain in 1910, Peñafiel resigned from the DGE due to health problems. He left before analysing the results of the 1910 census, the first of the twentieth century. Peñafiel left his post in a tense political atmosphere marked by the Plan of San Luis and the campaign led by Francisco I. The revolution that forced President Díaz to step down as president began with Madero in 1911.

Peñafiel's resignation from the DGE opened a period of criticisms and instability. He was succeeded by at least three other directors, and the agency's activities were interrupted from 1915 until 1922, after the revolution ended. During those years, the incompleteness of official statistics was a key object of concern for Peñafiel's successors. Although all of them questioned the production of statistics during the *Porfiriato*, they failed to produce any changes in the ways of doing statistics. On the contrary, the activities of the DGE's post-revolutionary officials only served to reinforce the techniques and practices inherited from their predecessors. The social relations which sustained them remained unmodified as well.

Albino R. Nuncio (1911–1914) and Salvador Echegaray (1912–1913) were two prominent successors of Peñafiel, both engineers with ample experience in administrative tasks. Trained in the army, Nuncio studied engineering at Lehigh University in the United States and worked for the Ministry of Development at various times between the Mexico City census of 1890 and 1914. He was assigned as a representative at the Universal Exhibitions, coinciding with Peñafiel in Paris in 1899 and in St. Louis in 1903. Echegaray, meanwhile, studied topography and mathematics, and was employed in the economic statistics section of the Treasury from 1897 to 1901 (INEGI 2010, 36).

In 1911, during Madero's government, Echegaray decided to transform the DGE's publications. The *Anuarios Estadísticos*, published since 1894, were replaced by the *Boletín de la Dirección General de Estadística*, whose first issue appeared in 1912. In that volume, Echegaray published the essay *Reorganización de las estadísticas. Informe al Secretario de Fomento por la DGE*, which unleashed a series of criticism towards the inconsistencies and inaccuracies of the DGE's earlier works. "[I]t seems unnecessary to make a detailed critique of the data that this office has published over the long years of its existence. Suffice to say, coping with the truth that science demands, that the outputs produced by this office have been untrustworthy and untimely" (Echegaray 1912, 1, author's translation).

The problem with the DGE's "notoriously false figures" had to do with the agency's bureaucratic officials. According to Echegaray, the lack of formal training of the individuals who collected the data and those in charge of interpreting them, especially when it came to doing further calculations, accounted for the weaknesses of the previous statistical inquiries. But far from proposing a new structure, he simply reproduced the organisational chain that had existed since the *Porfiriato* and that distinguished between the "simple" data collectors, whom he referred to as "statistical instruments", and the "observers" or the civil servants, who made estimates and calculations (Echegaray 1913, 4). This distinction and categorisation of the staff reflects a central issue: the need to ensure the existence of verification methods for statistics similar to "the measuring instruments used in scientific observations". In Echegaray's view, "statistical instruments" were "less valuable" than scientific tools, because "the knowledge, the intellectual qualities of that which I have called statistical instruments, are of relatively little importance" (Echegaray 1913, 6, author's translation). Hierarchies of this kind concealed a way of conceiving how the knowledge of official statistics was produced. For Echegaray, objectivity was a person's quality, just like the propensity to lie or manipulate data voluntary. Producing valid statistics demanded that the DGE watch over the individuals that made calculations and interpretations of the statistical numbers. To control them, he proposed fostering an environment of honesty and good faith, adding that he, as head of the DGE, would closely supervise their work. Echegaray was familiar with Poisson's law of large numbers and recognised that another way of ensuring certainty consisted in increasing the number of observations. However, he opted to concentrate on the personal supervision of officials (Echegaray 1913, 3).

The categories of population and economic activities that the bureaucracy of the *Porfiriato* had handled and displayed survived the revolutionary turmoil. Indeed, the content of the charts of the DGE's *Boletines* remained unchanged until 1922. These were works of the *Porfiriato* and mirrored the regime's version of the progress of the Mexican economy and its population. After the revolution, they continued to appear in the same form.

In 1922, during the government of Álvaro Obregón, the DGE was transformed into the DNE. Since it was turned into an office of the federal executive branch of power, the statistical agency no longer belonged to the Ministry of Development. The DNE resumed the tasks of producing national statistics in the way invented by the nineteenth-century bureaucrats. Despite the revolution, the daily work continued to be anchored in counting technologies and ways of gathering data proper to the hobby of collecting. Until at least the 1930s, the statistical instruments used by the DNE's bureaucracy simply reproduced ways of doing statistics based on what we might call 'personal ambiguity'. But by that time, the gentlemen of the *Porfiriato* had disappeared. What persisted was a social organisation in which citizens and civil servants lived amid tensions. Some smooth regions persisted, which not coincidentally were those inhabited by the poor, the indigenous peoples, peasants, and all who resisted being striated by a nation obsessed with standardisation. Civil servants at the DNE and the Ministry of Development grew in number, became specialists, and produced concrete outputs in the form of national statistics and censuses. In practice, however, they kept reproducing a social organisation in which the political legitimacy of statistical data rested upon personal qualities. The system was still relatively far from the increasing complexities of the statistical discipline and the forging of mechanical instruments for creating and verifying scientific data.

Final Ideas

Mexico's post-revolutionary governments, beginning with Venustiano Carranza (1914–1920), inherited from the *Porfiriato* an expanding public administration still loyal to existing hierarchies and to the tasks and responsibilities assigned to their posts. Scholars, usually inspired by Max Weber, identify bureaucratisation with processes of rationalisation in modern states. Weber included statistics in that process, depicting them as fair and objective tools of the state (Weber 1983, 1060; De la Peña and Wilkie, 1994). But in Mexico, even after the revolution of 1910, statistical institutions continued to be organised under patrimonial premises. In fact, the DGE seemed to assume that its tools were neither the forms nor the calculations, but the people. Consequently, it generated techniques for computing, classifying, and describing data based on a highly *personal* system, in which the only way for verifying the reliability of the information was the supervision and authority of the civil servants in charge. This reinforced an administrative organisation based on personal negotiations, without any mechanical means for checking or verifying its operations.

More than viewing the Mexican case as an exception or an error in the script of modernity, my goal was to re-think the meanings of quantification in late nineteenth-century and early twentieth-century Mexico. Certainly, this chapter does not discuss the processes of the rationalisation of public life. Rather, it is an invitation to inquire into the characteristics of statistical measurements. We must not assume that the quantifications associated with the bureaucratisation of public life are unproblematic or transparent simply because they are numerical in nature (Porter 2009). What I want to underscore is how the human factor, characterised by the primacy of the personal and personal relationships, was an integral feature of the statistical practices and the censuses and economic measurements produced in this period. The employees modified the calculations, magnitudes, and numbers that were produced.

What characterised the quantification of social life in Mexico from the late nineteenth to the early twentieth century was not spontaneity, the force of a supposed state rationality, or statistical information technologies. In the case of the DGE, national statistics were produced through the knowledges of neighbouring disciplines and thanks to the acquisition techniques used by collectors, especially those interested in national history. Thus, what we might call 'modern' statistical measuring was created by coupling practices of observation and knowledge, including the hobby of collecting, surveys, and censuses. Likewise, graphic representations, one of the great sophistications of Peñafiel's time, created routines, bureaucratic information technologies, and, we could say, technologies for producing statistical knowledge.

The processes analysed here are not simply anecdotal, for they refer to a time when archaeological objects were genuinely as important as surveys and censuses. In the end, the activities of both fields—the collection of archaeological artefacts and the accumulation of statistical data—referred to the 'Mexican population'. By transforming the population into a numerical entity, the Porfirian government sought to ease the cracks and splits denounced by those who believed that Mexico was divided, as it was, between white and indigenous peoples. Both types of practices, archaeological and statistical, produced figures and magnitudes, charts, and graphs that spoke of Mexico as a *mestizo* nation. The statistical representations on display at several universal exhibitions and at the National Museum gave statistics the character of a national patrimony, similar to the country's indigenous past. Statistics and their graphs, like antiquities, were erected as objects to be exhibited nationally and internationally by the Díaz regime and the post-revolutionary governments that followed (Garrigan 2012).

The statistical measurements produced by the DGE—which served as artefacts for political administration and the elaboration of policies—emerged from unexpected practices of knowledge that were the result of encounters between 'unexpected' citizens and bureaucrats who, far from being experts, behaved as gentlemen. These features invite us to think more deeply about the conditions of possibility of statistics and measurements considering the relevance of subjective factors.

References

Acevedo Rodrigo, Adriana and Paula López Caballero (coords). 2012. *Ciudadanos inesperados. Espacios para la construcción de ciudadanía ayer y hoy*. Mexico: COLMEX-CINVESTAV.

Agar, Jon. 2003. *The Government Machine. A Revolutionary History of the Computer*. Cambridge, Massachusetts: The MIT Press. Edition Kindle. https://doi.org/10.7551/mitpress/3336.001.0001.

Anderson, Benedict. 1993. *Comunidades Imaginadas. Reflexiones sobre el origen y difusión del nacionalismo*. Mexico: FCE.

Arrioja Díaz Viruell, Luis Alberto (ed.). 2016. *Registrar e imaginar la nación. La estadística durante la primera mitad del siglo XIX. Vol. I.* Mexico: El Colegio de Michoacán-Universidad Veracruzana-El Colegio de Sonora.

Azuela, Luz Fernanda. 1996. *Tres sociedades científicas en el porfiriato. Las disciplinas, las instituciones y las relaciones entre la ciencia y el poder.* Mexico: Sociedad Mexicana de Historia de la Ciencia y la Tecnología.

Blanco, Mireya, and Moncada José Omar. 2011. "El ministerio de fomento, impulsor del estudio y del reconocimiento del territorio mexicano (1877–1898)." *Investigaciones Geográficas* 74: 74–91.

Cházaro García, Laura. 2000. Medir y valorar los cuerpos de una nación. Un ensayo sobre la estadística médica en el siglo XIX. Mexico: Ph.D. Thesis, UNAM.

———. 2018. "Médecins, statistique et recensement: Les raisons d'une entente cordiale." *Histoire et Mesure* 33 (2): 33–60. https://doi.org/10.4000/histoiremesure.7970.

Cosío Villegas, Daniel. 1985. *Historia moderna de México. El porfiriato. La vida política interior. Segunda parte.* Mexico: Editorial Hermes.

Cuevas Cardona, Consuelo. 2006. *La investigación biológica y sus instituciones en México entre 1868–1929.* Mexico: Ph.D. Thesis, UNAM.

Daniel, Claudia J. 2014. "Imágenes estadísticas en la cultura visual de la Argentina moderna." *Caiana. Revista de Historia del Arte y Cultura Visual del Centro Argentino de Investigadores del Arte (CAIA)* 5: 80–94. http://caiana.caia.org.ar/template/caiana.php?pag=articles/article_2.php&obj=157&vol=5. Accessed 28 September 2020.

De la Peña, Sergio, and James Wilkie. 1994. *La estadística económica en México. Los orígenes.* Mexico: Siglo XXI and Universidad Autónoma Metropolitana.

De Mier, Sebastian B. 1900. *México en la exposición universal internacional de Paris.* Paris: Imprenta de J. Demoulin.

Deleuze, Gilles, and Félix Guattari. 1980. *Capitalisme et Schizophrénie. Mille Plateaux.* Paris: Les Éditions de Minuit.

Desrosières, Alain. 2010. *La politique des Grands Nombres. Histoire de la raison statistique.* Paris: Éditions La Découverte. https://doi.org/10.3917/dec.desro.2010.01

Dirección General de Estadística (DGE). 1894. *Anuario Estadístico de la República Mexicana 1893.* Mexico: Dirección General de Estadística, Oficina Tipográfica de la Secretaría de Fomento.

———. 1897. *Censo General de la república Mexicana Verificado el 20 de octubre de 1895.* Mexico: Oficina Tipográfica de la Secretaría de Fomento.

———. 1898. *División municipal de la República Mexicana. Cuarta edición corregida y aumentada.* Mexico: Oficina Tipográfica de la Secretaría de Fomento.

———. 1901. *Anuario Estadístico de la República Mexicana 1900.* Mexico: Dirección General de Estadística, Oficina Tipográfica de la Secretaría de Fomento.

Echegaray, Salvador. 1912. "Informe presentado al Secretario de Fomento por el Director General de Estadística." *Boletín de la Dirección General de Estadística* 1: 1–10.

———. 1913. "El método en estadística y algunas de sus aplicaciones a la estadística demográfica en México." *Boletín de la Dirección General de Estadísica* 2: 1–12

Garrigan, Shelley E. 2012. *Museums, Monuments, and the Creation of National Identity. Collection Mexico.* Minneapolis and London: University of Minnesota Press. https://doi.org/10.5749/minnesota/9780816670925.001.0001.

Hacking, Ian. 1986. "Making Up People." In *Reconstructing Individualism: Autonomy, Individuality, and the Self in Western Thought*, edited by Thomas C. Heller, Morton Sosna, and David E. Wellbery. Stanford, Stanford University Press.

Hale, Charles A. 1991. *La transformación del liberalismo en México a fines del siglo XIX.* Mexico: Vuelta.

Hermosa, Jesús. 1870. *Compendio elemental de Geografía y Estadística de la República Mejicana. Escritos con el título de Manual.* Mexico and Paris: Librería de Rosa y Bouret.

Instituto Nacional de Estadística, Geografía e Informática (INEGI). 2008. *Antecedentes de la Estadística en México.* Mexico: INEGI.

———. 2010. *125 años de la Dirección General de Estadística, 1882–2007.* Mexico: Colección Memoria-INEGI.

Liceaga, Eduardo. 1892. *Censo de la Ciudad de México. Ensayo para ejecutar esta operación, presentado a la Asociación Americana de Salubridad Pública.* Mexico: Imprenta del Gobierno Federal.

Lozano, María. 1991. *La Sociedad Mexicana de Geografía y Estadística (1833–1867). Un Estudio de caso: La Estadística.* Mexico, M.A. Thesis, Universidad Nacional Autonoma de Mexico.

Mayer Celis, Leticia. 1999. *Entre el infierno de una realidad y el cielo de un imaginario: Estadística y comunidad científica en el México de la primera mitad del siglo XIX*. México D.F.: El Colegio de México, Centro de Estudios Históricos. https://doi.org/10.2307/j.ctvhn0d1g.

———. 2003. *La tan buscada modernidad científica. Boletín del Instituto Nacional de Geografía y Estadística de 1839.* Mexico: IIMAS-UNAM.

Medeles, Ana. 2011. "Medición y población en México a finales del siglo XIX: Estadísticas electorales". M.A. Thesis, Universidad Nacional Autónoma de México.

Medeles, Ana. 2016. "Documento histórico. La ley del 26 de mayo de 1882 que constituyó la Dirección General de Estadística." *Estadística y Sociedad*

4: 153–159. https://seer.ufrgs.br/estatisticaesociedade/article/view/64424. Accessed 27 May 2021.

Müller-Willie, Staffan. 2017. "Names and Numbers: 'Data' in Classical Natural History. 1758–1859." *Osiris* 32: 109–128. https://doi.org/10.1086/693560.

Pacheco, Carlos. 1882. *Memoria de la Secretaría de Fomento, Colonización, Industria y Comercio*. Mexico: Imprenta de Fomento.

Palti, Elías José. 2005. *La invención de una legitimidad. Razón y retórica en el pensamiento mexicano del siglo XIX. Un estudio sobre las formas del discurso político*. Mexico: FCE.

Patriarca, Silvana. 1996. *Numbers and Nationhood. Writing Statistics in Nineteenth Century Italy*. Cambridge: Cambridge University Press. https://doi.org/10.1017/CBO9780511523458.

Peñafiel, Antonio. 1882. "Clasificación de las materias que forman la Estadística General." In *Memoria de la Secretaría de Fomento, Colonización, Industria y Comercio*, 125–179. Mexico: Oficina Tipográfica de la Secretaria de Fomento.

———. 1883. "Ensayo de análisis estadístico sobre lesiones." *Gaceta Médica de México* 7 (18): 113–131.

———. 1884. "Estadística general." *Periódico oficial de la Dirección General de Estadística*. 15 December.

———. 1889a. *Boletín Semestral de la Estadística de la república Mexicana*, No. 2. Mexico: Oficina Tipográfica de Fomento.

———. 1889b. *Explication de l'Édifice Mexicain a l'Exposition Internationale de Paris en 1889*. Barcelone: Imprimerie d'Espasa et Cie.

———. 1890a. *Estadística general de la República Méxicana*. Mexico: Secretaría de Fomento.

———. 1890b. *Monumentos del arte mexicano antiguo: Ornamentación, mitología, tributos y monumento*. Berlin: A. Ascher and Co.

———. 1895. *El lago de Texcoco*. Mexico: Secretaría de Fomento.

———. 1897. *Nomenclatura geográfica de México. Etimologías de los nombres correspondientes a los principales idiomas que se hablan en la República*. Mexico: Imprenta Tipográfica de la Secretaría de Fomento.

———. 1897–1903. *Colección de documentos para la historia mexicana, 6 volúmenes*. Mexico: Oficina Tipográfica de la Secretaría de Fomento.

———. 1901. *Diccionario de la lengua tarasca o de Michoacán por Fr. Maturino Gilberti*. Mexico: Tipografía de la Oficina Impresora Estampillas.

Pérez Gallardo, Basilio. 1873. *Cuadro estadístico de la división territorial de la República Mexicana*. México, Imprenta del Gobierno en Palacio.

Peust, Otto. 1910. *Estadística Agrícola. Razones que determinan la naturaleza de los datos más urgentes para fundamento de la política agraria e instrucciones sobre la manera de reunir datos*. Mexico: Imprenta y Fitotipia de la Secretaría de Fomento.

Porter, Theodore. 1994. "Making Things Quantitative." *Science in Context* 7 (3): 389–407. https://doi.org/10.1017/s0269889700001757.

———. 1995. *Trust in Numbers. The Pursuit of Objectivity in Science and Public Life*. Princeton New Jersey: Princeton University Press. https://doi.org/10.23943/princeton/9780691208411.001.0001.

———. 2009. "La estadística y el curso de la razón pública: El compromiso e imparcialidad en un mundo cuantificado". *Empiria. Revista de Metodología de las ciencias sociales* 18, July–December: 19–35.

Poovey, Mary. 1998. *A History of the Modern Fact. Problems of Knowledge in the Sciences of Wealth and Society*. Chicago: The University of Chicago Press. https://doi.org/10.7208/chicago/9780226675183.001.0001.

Prévost, Jean-Guy, and Beaud, Jean-Pierre. 2012. *Statistics, Public Debate and the State (1800–1945): A Social, Political and Intellectual History of Numbers*. London: Pickering and Chatto. https://doi.org/10.4324/9781315655161.

Rodríguez-Kuri, Ariel. 1996. *La experiencia olvidada. El ayuntamiento de México: Política y gobierno, 1876–1912*. Mexico: El Colegio de México-UAM-A.

Rodríguez de Romo, Ana Cecilia, Gabriela Castañeda, and Rita Robles Valencia. 2008. *Protagonistas de la medicina científica mexicana. 1800–2006*. Mexico: UNAM and Plaza Valdés Editores.

Sánchez-OEconomo, Esteban. 2018. *L'image de prospérité nationale du Mexique à l'Exposition universelle de Paris 1900, Mémoire de Master 2*. France: École des hautes études en sciences sociales, Centre Alexandre-Koyré.

Tenorio-Trillo, Mauricio. 1998. *Artilugio de la nación moderna. México en las exposiciones universales, 1880–1930*. Mexico: Fondo de Cultura Económica.

Tufte, Edward R. 2007. *The Visual Display of Quantitative Information*. Cheshire, Connecticut: Graphics Press.

Urquiza, Humberto. 2020. El Instituto Nacional de Estadística, Geografía e Informática: A History of Innovation Rooted in Tradition. In *Oxford Research Encyclopedia of Latin American History*.

Von Oertzen, Christine. 2017. Machineries of Data Power: Manual Versus Mechanical Census Compilation in Nineteenth Century Europe. *Osiris* 32: 129–150. https://www.journals.uchicago.edu/doi/10.1086/69391.

Weber, Max. 1983. *Economía y Sociedad*. Mexico: FCE.

Zuleta, María Cecilia. 2000. "La Secretaria de Fomento Agrícola en México, 1876–1910: La invención de una agricultura próspera que no fue." *Mundo agrario. Revista de estudios rurales* 1 (1): s/n.

CHAPTER 5

Ethnic Origin, Race, and Nation in the Argentine Censuses, 1869–1914

Hernán Otero

Throughout the nineteenth century both the Argentine nation and the Argentine state were made. Rightly so, the studies on the making of the Argentine nation prioritised the role played by intellectuals in developing different ideas of the nation (Halperín Donghi 1980). They also enquired on the tools used by the state to carry out a task that became even more imperative as rapid economic growth and the massive arrival of Europeans blurred the social and ethnic features established during the

This chapter is an updated version of the article "Estadística censal y construcción de la Nación. El caso argentino, 1869–1914," *Boletín del Instituto de Historia Argentina y Americana Dr. Emilio Ravignani*, n° 16–17, 1998, 123–149.

H. Otero (✉)
Instituto de Geografía, Historia y Ciencias Sociales, IGEHCS, CONICET–UNCPBA, Tandil, Argentina
e-mail: hernan.otero@conicet.gov.ar

C. T. Lanata-Briones et al. (eds.), *Socio-political Histories of Latin American Statistics*, Studies of the Americas,
https://doi.org/10.1007/978-3-030-87714-9_5

colonial period. Among those tools, the development of the educational system, the creation of a patriotic rhetoric through symbolic festivals and commemorations (Bertoni 2001), and compulsory military service stood out. State action was not reduced to these tools since, to varying degrees, all administrative layers contributed to the cultural movement behind the making of the nation (Terán 1987) that positivism embodied as the dominant philosophy from 1880 onward. In the case of statistics, however, this purpose had its roots in longer-standing political and scientific conceptions.

This chapter aims to understand the role played by the national statistical system in the building of the Argentine nation, through the analysis of a privileged instrument of observation: national population censuses. The starting premise, verified for other contexts, is that national statistical apparatuses define cognitive and discursive matrices that play an important symbolic role in the creation and diffusion of a certain image of society and of the nation. Naturally, the symbolic creation of the nation generated by the censuses exerted less influence than mechanisms such as the educational system, which acted directly on people's identity. This limitation is reduced, however, when considering the long-term effects of the censuses on the way of measuring and thinking about the Argentine social process.

In response to multiple demands for information, one of the fundamental changes that European states experienced during the nineteenth century was the development of universal and centralised statistical systems. Economic, demographic, educational, and criminal statistics, among others, provided the state with tools to guide its policies with a marked optimism. Sporadic uncertainties concerning the possibility of a scientific exercise of the state's powers still existed, however. In this context, the statistical system played a fundamental role in the image that states portrayed of themselves and of the dynamic and heterogeneous populations that lived within their borders. This characteristic was not unique to the Argentine case, of course. It constituted a generalised feature of the period, as the cases of France, Italy, and the United States show, to name only the countries that exerted the greatest influence on the Argentine census experience.[1] Statistics became, since then,

[1] See Hervé Le Bras (1988), Dora Marucco (1996), Silvana Patriarca (1996), Margo Anderson (1988), and Schor (2009) for further information on these countries. Regarding the role of censuses as makers of an image of the nation in Latin America see Hernán

a formidable tool in the development of a general representation of the people and the nation through the double operation of collective self-identification within each country and of comparison to/differentiation with others.

The nation is an intellectual construct developed using a set of symbols and beliefs from a partly real and partly imaginary story, but also using evidence from specific populations. For this reason, demography and statistics became, especially during the second half of the long nineteenth century, crucial intellectual fields to define the nation, given their ability to determine 'the criteria of belonging' to enumerate individuals. These criteria were numerous, especially in countries like Argentina in which the pre-existence of indigenous peoples, the forced integration of Afro-descendants, and the avalanche of migrants gave rise to a dynamic and heterogeneous demographic ensemble.[2] Thus, it is essential to examine the role that the census assigned to nationality, ethnic origins, and race.

The required starting point of this research is the assertion, common in the Hispanic-American context, that nineteenth-century statistical enquiries omitted from national censuses questions regarding the 'colour' of people, which could refer to the disgraceful ethnic and racial distinctions of the colonial past.[3] Following this thesis, the rejection of a polychromatic palette was not only the product of growing miscegenation between ethnic groups. It was, above all, the result of the change in the perception of society given the emergence of liberalism and the principle of equality under the law embodied in the constitution. Concerning statistics, this principle implied the revolutionary translation of a socio-ethnic grid inherited from the three-estate tradition of *Ancien Régime* societies to a reading based on the universal principle of citizenship, as well as the adoption of nationality as the central criterion that defined individuals. The use of individuals as the main, although not exclusive,

Otero (2018). The supremacy of imperial powers gives particular interest to Southeast Asia and colonial India, analysed by the influential text by Benedict Anderson (1983) and Arjun Appadurai (1996), respectively.

[2] In Argentina at the time, indigenous populations were referred to as Indians, while the word black instead of Afro-Argentine was used. These terms are employed in this chapter to reflect the perceptions of the period under analysis.

[3] For example, Gonzalo Aguirre Beltrán (1970) and Nicolás Sánchez Albornoz (1994). It is worth noting that the argument behind not using socio-ethnic categories due to their discriminatory connotations was not put forward by the census takers.

unit of analysis and the almost lack of attention given to illegitimacy point in the same direction. However, the amendment made in the census questionnaires under the influence of liberal principles did not translate into equal consideration of the demographic components that were to constitute the new nation. On the contrary, the adoption of a non-discriminatory grid at the statistical level (data collection and quantification) coexisted with a hierarchical and evolutionary reading at the discursive level (comments and interpretations of the data). Based on these considerations, this chapter analyses the treatment that the statistical apparatus gave to the groups that made up the Argentine population (Indians, blacks, and immigrants). The joint analysis of these three populations, following the choice of the census-takers themselves, makes it possible to detect basic features that are more difficult to perceive when they are analysed separately. In a similar sense, a comparison with the United States and Canada is included in order to understand the Argentine case more accurately. These two countries had, at the time, a greater degree of statistical ethnicisation and were a reference used by the census -takers themselves, specifically the United States.[4]

This analysis focuses on the categories used and their consequences at the interpretive level. It is based on a simplified version of the concept of statistical chain (Merllié 1989), retaining three main components: the relationship between the enumerator and the person being registered at the time of the census; the choice of categories and the making of cross tabulations and charts; and the final discursive interpretation of the data. Although references are made to the first link, the weight of the argument falls on the last two components.

The objects of study are the first three Argentine national population censuses of 1869, 1895, and 1914 in which, beyond specific differences, there is a conceptual continuity. The official nature of the censuses, the standardised and 'conservative' imprint that is typical of any statistical system, and the continuity of technical teams allow an overall analysis of a period that was characterised by a strong "ideological consensus" (Hale 1991).[5] An analysis of the statisticians as a political-intellectual field could show greater changes than those presented here. However, for the

[4] For a comparison with other Latin American countries see Mara Loveman (2014).

[5] The 'conservative tendency' of statistical systems derives, according to Joshua Goldstein (1990), from the need to obtain comparable results over time, making it difficult to incorporate new topics and forms of measurement. The technical staff who designed the

reasons mentioned, these changes were not transferred to the censuses, which emphasises their paradigmatic continuity. Another starting premise is that census ideology is not limited to rational and volitional aspects as it focuses on the resulting matrices regardless of the goals consciously pursued by the social actors.

During the second half of the long nineteenth century, a series of decisive changes gave rise to the so-called modern Argentina. The creation of an institutional framework with the enactment of the constitution in 1852 allowed the progressive consolidation of the Argentine state, especially after the arrival of General Julio A. Roca (1843–1914) to the presidency in 1880. Argentina integrated itself economically into the international market as an exporter of primary goods and as a recipient of foreign investment, particularly of British origin. This integration was considered necessary for the consolidation of the state and for modernisation. From the 1880s, the territorial expansion incorporated to the nation the vast territories occupied by the indigenous peoples in Patagonia and in the northeast of the country. A massive influx of European immigrants contributed to the redefinition of the socio-ethnic and economic profile of the country, especially in the Central-Littoral region, in the context of rapid population growth.[6]

A constitutive part of the state-building process was the progressive emergence of a statistical system following the creation of the Office of National Statistics (*Oficina de Estadística Nacional*, 1864–1875) and, above all, of the General Bureau of Statistics (*Dirección General de Estadística de la Nación*, 1894–1943). Although population censuses were not carried out at ten-year intervals as the constitution stipulated until 1960, the three censuses of the period defined the long-term intellectual matrix in terms of ethnic categories. This matrix was modified with

national population censuses of the period included Diego G. De la Fuente (superintendent of the first national census, director of the 1881 census of the Province of Buenos Aires, and president of the 1895 Census Commission); Gabriel Carrasco (director of the 1887 census of the Province of Santa Fe, and member of the 1895 Census Commission); Alberto B. Martínez (member of the commission of the 1887 census of the City of Buenos Aires, director of the 1890 census of the Province of Córdoba, member of the 1895 Census Commission, and president of the 1914 Census Commission); and Francisco Latzina (director of the 1887 census of the City of Buenos Aires, and member of the 1914 Census Commission). On the range of activities carried out by the census takers, see Otero (2006), Claudia Daniel (2011), and Hernán González Bollo (2014).

[6] The population of the country quadrupled between 1869 and 1914.

the censuses of 2001 and 2010 that reincorporated the measurement of indigenous and Afro-descendant populations, respectively. As this chapter shows, the period 1869–1914 contains long-lasting tools for the study of both the Argentine population and the categories through which the population becomes intelligible.

For a better development of the chapter's objectives, the following three sections analyse separately the enumeration of the indigenous, black, and European populations, respectively. Finally, the conclusion makes explicit the advantages derived from the joint analysis of these three populations and reveals the cognitive matrix of the national censuses of the period and its long-term symbolic effects for the construction of a certain image of the Argentine nation.

The Weakest Enemy

> The old Indian issue is not an issue of Indians, it is an issue that relates to the DESERT (sic). The Argentine Indian, by himself, is perhaps the weakest and least fearsome enemy of civilisation; barbarian, superstitious, vicious, naked, he even has an enemy in his weapon. Eliminate him altogether, but leave the desert, and you will immediately have two hundred *gauchos* take his place and replace him, [*gauchos*] too many and daring to threaten the borders of half the states and to have four or five thousand veterans under control. And on the contrary: eliminate the desert; this desert that intrudes and comprises us, linking us almost to the shores of the cities, and the Indian, like the *montonero*, will disappear without further effort. Diego De la Fuente (Argentina 1872, LIV–LV, author's translation)

Together with the population of colour and European immigrants, indigenous peoples constitute one of the basic components of Argentine demographic history. Seen as a whole, the census framework determined a reading of the Argentine social process that enhanced the role played by immigrants while dissolving the presence of Indians and blacks. The physical elimination of indigenous peoples after General Roca's Conquest of the Desert (*Campaña al Desierto*) in 1879 and the Conquest of the Chaco (*Conquista del Chaco*) in 1884 was preceded and accompanied by an operation of statistical disappearance. This statistical disappearance was due to the omission of questions on race and colour, typical of the colonial grid, and of the dominant ideological orientations within the intellectual elite and the census system.

During the nineteenth century, a huge share of today's Argentine territory was outside the control of the state, which maintained a mobile and conflictive border with the non-dominated Indian tribes. Although surveying these populations while at war with the 'white population' was beyond the real possibilities of the state, the Indians were counted by the census-takers in 1869 and 1895 and were partially registered in 1914. Beyond its conjectural nature, the censuses explicitly sought to estimate the militarily active indigenous population. The strategic value of this information was reinforced by the fact that, unlike the data on the rest of the population, it was gathered by the commanders and military chiefs stationed at the country's borders. Thus, from the technical point of view, the enquires of 1869 and 1895 were censuses, strictly speaking, only for the population (mainly but not exclusively white) of the territories controlled by the state, while they were estimates regarding the population (mainly but not exclusively indigenous) of uncontrolled territories. The situation persisted, albeit to a much lesser degree, in 1914.

These differences were not exclusively due to the—admittedly enormous—technical difficulties of counting the 'weakest enemy'. They also derived from the notions about the Indians that prevailed within the white population. The composition by age and gender of the Indian population constitutes a paradigmatic example in this regard. A description of the Indians of the province of La Pampa, for example, suggested a pattern that with slight variations was found within all indigenous groups: "chiefs Cañumil and Juanpicliun [have in the] first [case] 150 spears, or men of war, which suggests a total of 750 individuals; [in] the second [case] 160 spears or men of war, thus about 800 individuals. Within this this group there are 1550 inhabitants" (Argentina 1872, 613, author's translation).[7] The Indians, who were not distinguished based on their gender nor age, constituted an undifferentiated group, identified only by their military capacity and their status as real or potential enemies.

Other aspects of the presence of indigenous peoples in the censuses are equally relevant. As in the United States, there was no explicit definition of said population. Contrary to a socio-cultural definition whatever

[7] Although De la Fuente did not indicate the method for calculating the indigenous population, it is evident that it results from multiplying the number of men of war by a constant factor of 4 or 5 people. By 1914, the census-takers considered that the indigenous population figures from the first two censuses were much lower than reality or even simply "arbitrary" (Argentina 1916, I, 67; IV, 501, author's translation).

it would have been, the concept of Indian had an exclusively legal and, *a fortiori*, essentially political connotation for census-takers. Within the census logic, an Indian was someone who lived in the 'desert', a term used at the time to refer to the uncontrolled territories. The association between the terms 'Indian', 'gaucho', and 'montonero' proposed by De la Fuente illustrates the political-territorial basis of statistics. This restrictive delimitation was not only a legalistic description of Argentine society.[8] It was also a logical extension of the classical Domingo Faustino Sarmiento's opposition between "civilisation and barbarism" (*Civilización y barbarie. Vida de Juan Facundo Quiroga*, 1845) and of the distinction between peoples according to levels of civilisation, initiated by the romanticists and continued by positivists.

The association with the desert was the central element behind the definition of the indigenous population, but not the only one. Religion was another way of classifying them, as shown by the census tables that reinforced the legal classification through the dichotomy 'Christians and Indians', reminiscent of the ethnocentric tradition of opposition / exclusion. Religion, however, was at the time a secondary aspect in statistics, which were defined by liberal and secular principles. The marginal nature of the religious criterion also explains why, once the indigenous territories were controlled by the state, the statistical apparatus did not attempt to understand the distinction between evangelised and non-evangelised Indians, despite that the question on religion was asked to everyone in the second census. Indeed, this question was not an attempt to approach the indigenous problem, since the prescriptive nature of the census forms meant that the Argentine inhabitants were considered Catholics.[9]

Following this reasoning, it is not surprising that the polarisation generated around the subjugation of the indigenous peoples was associated with their mobility and residence, in a logic in which sedentism and nomadism reinforced the initial opposition between civilisation and barbarism. This is clear, for example, in the questioning of the estimates

[8] Legalism is defined as the measurement process based on the legal categories established by the state and not on the observed social practices. This trend had important effects, for example, on the measurement of marital status (Otero 2006).

[9] "As almost the entire Argentine population is Catholic, the census-taker will only ask the question about religion when he has reason to believe that the individual is not Catholic" (Argentina 1898, II, CXXI, author's translation). The conversion of Indians to Catholicism established by the constitution could not but reinforce this trend.

of the 1865 census of the province of Salta in which "the wild population, which lacked safe or legal residence, was added to the civilised population who was under provincial jurisdiction and had a legal and fixed residence. Some 20,000 individuals lived in such [wild] conditions and were distributed proportionally among the population of the departments". The author of the first census concluded that "discarding such population" the province of Salta had 84,000 inhabitants (Argentina 1872, 534, author's translation). Hence, 'population with legal and fixed residence' meant civilised population and 'population which lacked safe and legal residence' referred to the Indian population. Moreover, sedentary and conquered Indians were not seen as Indians.

The same idea, reinforced by the ideological role that the statistical apparatus assigned to property, reappeared in the second census through an argument that paradigmatically linked sedentism, land ownership, and sense of nationhood. The Indian, who was mobile and hence incapable of owning property like white people could, lacked national sentiments. In a society defined by natural law and reinforced by liberalism and positivism where ownership was a basic feature of civilisation and patriotism, non-owners could only be savages. The inertia of the argument persists: the obvious continuity between 'love for the land' and 'love for the country' transformed indigenous peoples into potential 'traitors'. In De la Fuente's words:

> The easy acquisition of land is undoubtedly one of the greatest incentives, not only to attract population, but especially to make it stay in a particular place and make it civilised and prosperous. The Arab, a nomad who places his tent in the desert, is ready to take it apart as soon as he exhausts the resources he can obtain around him; and the wild Indian of the Pampas or of the sub-tropical forests, has no affection for the land and almost absolutely lacks that noble feeling called love for the country and which is developed and consolidated by the encouragement of land ownership. (Argentina 1898, II, CXIV, author's translation)

The enormous technical difficulties associated with the accounting of indigenous peoples, such as the marked rejection to census taking especially in the areas of recent military occupation, enhanced the preceding arguments.[10] In the census records (the first component in the statistical

[10] On the causes of why indigenous peoples rejected the census, especially due to the fear of being deprived of their children and women, see Argentina (1916, I, 461–463).

chain) self-denomination was not infrequent among Indians. However, it was not widespread given the power asymmetry between enumerators and indigenous peoples, which led the latter to try to avoid being surveyed. This was not the situation in the oldest provinces of the Central-Littoral region and of the north-west of the country. In those provinces, the absence of an Indian population was due to the exclusion of the category from the records, which influenced the possibility of self-identification. While both logics acted together, removing the category was more decisive since it was a decision made in advance and independently of the technical problems of the enquiry. Moreover, these problems were not unconnected to the political will of the state. While the quantification of the material wealth of the nation reached levels of considerable precision, less effort was displayed towards population accounting. For census-takers it was evident that "only a special census of indigenous peoples, which will take just a few months of work with selected commissioners, locals and *baqueanos*, would provide a very approximate number of how many exist in the territory" (Argentina 1916, I, 465, author's translation). However, only in 1965 did the Argentine state undertake such operation with the National Indigenous Census (*Censo Indígena Nacional*), in a context more attentive to indigenous peoples' problems than one with combative nineteenth-century aims.

The change from the colonial census grid to the liberal census grid was common to other Latin American countries. However, in Argentina, where the indigenous population was smaller, it was instrumental in generating a homogeneous perception of the population and in establishing a partially artificial vision of society early on. Regardless of its egalitarian benefits, the new grid was functional to the non-recognition of the pre-existence of indigenous 'nations' and to a historical view that legitimised the rights of white people. According to this logic, indigenous peoples could only be considered as agglomerations of tribes or 'savages' and not as foreign nations, a legal status that would have called into question the rights of the state over the conquered lands.[11]

For a case study on the self-identification of individuals as indigenous in 1895 see Gabriela Nacach (2012).

[11] The issue was an important topic of debate as Sauze's thesis shows (L. Sauze, *Las tribus salvajes ante el derecho internacional*, Universidad de Buenos Aires, 1873, referenced in Biagini 1995, 69).

These developments explain the statistical disappearance of Indians in Argentina. Those who survived the military conquests and became part of the controlled territories, ceased to be Indians for the national statistical apparatus. A constant and artificial decline of the indigenous population took place, for which census-takers provided little explanation of. For the author of the second census, the drop that occurred during the first intercensal period from 93,138 people in 1869 to 30,000 in 1895 implied that the indigenous population had decreased "due to being confused with the civilised population or because the gaps that death leaves are not able to be filled by the new generations" (Argentina 1898, II, L, author's translation). The first explanation was very general as it included both the incorporation of the occupied territories and, eventually, miscegenation. The second reduced the problem to the relationship between procreation and mortality without differentiating the weight of each factor, especially the influence of the mortality produced by military actions.[12]

A Beautiful White Race

> The racial question, so important in the United States, does not exist in the Argentine Republic, where it will not take long for its population to be completely unified, forming a new and beautiful white race, product of all the European nations fertilised on the soil of the Americas. Gabriel Carrasco. (Argentina 1898, II, XLVIII, author's translation)

Like the indigenous peoples, the black population occupied a secondary place in the Argentine census discourse. Although its numerical importance was less than in Brazil, the United States, or the West Indies, its statistical isolation cannot be explained only through quantitative criteria. Rather, its absence was ideological. It was linked not only to the egalitarian motivations of the new liberal grid but also to the evolutionary and non-egalitarian conviction concerning the existence of a hierarchy within the groups that formed the nation.[13]

[12] International emigration was not contemplated. Naturally, the logic of the classification meant that Indians abroad were not Argentines, as were the Argentines abroad, who were estimated or counted in the three censuses (Otero 2006).

[13] The omission of the race category in 1869 was an important decision since colour was recorded by the Statistical Registry and by the censuses of both the city and the

Questions about the 'colour' of the population were absent due to technical reasons, subject to the principle of methodological minimalism which aimed to obtain data whose reliability and validity did not raise doubts:

> When the Executive Committee of the 1895 National Census discussed the census, the point related to investigating the ethnic composition of the country's population was carefully discussed, agreeing not to do it, first, because, given the small absolute and relative number of blacks, mulattos, and civilised Indians, the investigation was irrelevant; and second, taking into account that, except for the purebred blacks unable to escape the classification, the mulattos and Indians to a large extent would have been counted for as white, providing inaccurate and inferior figures of reality. (Argentina 1898, II, XLVI, author's translation)

Gabriel Carrasco's explanation pointed out two kinds of problems: the technical complexity of the census, especially in relation to self-definition, and the share of black people in the population. The latter argument was false around 1895, at least for the country's capital.[14] The difficulties concerning self-definition, however, were substantial, especially within the society surrounding the *Río de la Plata* that was characterised by a high degree of miscegenation and by certain upward social mobility that allowed blacks and mulattos to 'whiten' with relative ease.

Regardless of its racist ideological considerations, the census argument was based on the explicit recognition of the enormous difficulties that existed when measuring race and on the conviction that a significant part of the population of colour could not be classified as such by the external gaze of the census. These acknowledgments implied an outright denial of any essentialist idea about the concept of race. Indeed, the alternative to the decision to suppress the category would have entailed the conviction that race constitutes an objective attribute to the external gaze that is traceable through generations. On the contrary, census-takers were aware

province of Buenos Aires until the 1880s. Race was also accounted for in contemporary censuses in the United States and Brazil.

[14] With few exceptions (Uruguayans, Spanish, French, and Italians), the number of foreigners accounted for in the 1887 census of the City of Buenos Aires was smaller than that of the black population.

of the constructed and dynamic character of race and of the inconvenience of reducing it to a purely physical meaning. It is in these technical discussions where the most progressive and reformist aspects of the census discourse are found.[15]

However, the omission of questions about colour was not enough to transform the census discourse into a non-racist, egalitarian discourse. On the contrary, the national censuses of the second half of the nineteenth century had one of their fundamental pillars in the concept of race. Race was considered an explanatory factor of socio-demographic phenomena given its causal association with population density, public education, and potential levels of conflict in society. In all cases, arguments pivoted around the mixture of races. However, census-takers did not use the term 'melting pot' whose subsequent success in the Argentine imaginary and historiography is well known due to Gino Germani (1962), among others. The historical narrative based on the census discourse suggests that Argentina was characterised by a constant process of racial mixing, whose harmony and speed the census takers repeatedly emphasised. This mixing was first among whites, blacks, and Indians and later between natives and immigrants. The evaluation of these processes was always positive and, at least in a first reading, there were no differential connotations about the hierarchy of the groups that participated in the mix.

The positive character of the racial mixture was based on the principles of natural selection formulated by Charles Darwin in 1859, as summarised by Carrasco when commenting on the second census:

> A simple reading of the data [share of immigrants that entered the country by nationality] is enough to understand how great the ethnic influence of the foreign element in the Argentine Republic is, and that a new race, intelligent and vigorous, has been formed and continues to form, since following the laws of natural selection the products of the mix are superior to each of the beings that gave life to it. The fact found regarding superior animal species, that their products are improved by the interbreeding

[15] George Andrews (1989, 94, author's translation) argues that the disappearance of the black population was not the consequence of higher mortality but the result of several factors: the loss of their proportional weight in the face of the migratory avalanche and the "hypothesis of statistical transfers" since official sources would not have reflected "exactly the racial realities of the city." This would have made possible, according to the author, a whitening in two phases: self-identification of blacks as mixed-race to escape race prejudice followed by the transfer of the mixed-race population to the white category.

> of the species, has also been reproduced as applied to the human race wherever observations have been made. These theories have had an excellent confirmation in the Argentine Republic, where it is observed that new generations emerge due to the connection of European men with American women, and vice versa, stronger and more beautiful than the individuals of which they proceed from. (Argentina 1898, II, XLIII, author's translation)

Carrasco's quote constituted an argument that was repeated in the three national censuses, although it acquired greater force as the migratory inflow intensified from the 1880s. Ideas about the benefits of natural selection did not seem to imply, in principle, a value judgement on the intervening racial groups since the mixture generated a new entity, different and superior to the races that participated in it.

This egalitarianism was only apparent since various discursive mechanisms outlined an asymmetric image of the relative contribution of each racial group to the development of the new nation, an image that gave hierarchical and evolutionary connotations to the Argentine melting pot. A key mechanism was the constant association between the black condition and adverse socio-cultural factors. Thus, when in 1914 enumerators sought for a negative point of reference to compare the level of school attendance of the Argentine population, they found it logical to link the state of Alabama in the United States "populated by 2,183,093 inhabitants, of which 909,261 are black, with the rich and cultured Province of Buenos Aires inhabited by a population equal to that of the US state, where, only rarely, one finds an inhabitant of black complexion." Then, when looking for a positive point of reference, the census-taker mentioned the state of Minnesota "whose inhabitants are 99 percent white." This methodology used race as the central variable of the comparison without considering other intervening variables such as the place occupied by the black population within the US social structure. Based on it, the census-taker concluded that in the United States the "high proportion of 7.7% of illiterates relates to entire population of that great nation, in which there are various races, such as black, Indian, and Chinese, whose culture does not stand out" (Argentina 1916, I, 172–173, author's translation). In this logic, the negative condition of not attending school became an inherent trait of the black racial category. This corresponded to a hierarchical and asocial view of society.

Less elusive than previous associations, explicit considerations of enumerators left little room for second readings. In 1914, at a time when the mixture with Indians and blacks was well advanced and the presence of European immigrants reached its peak, Alberto B. Martínez—quoting verbatim the *Historia de Belgrano* by Bartolomé Mitre, one of the fathers of Argentine historiography and President between 1862 and 1868—argued that:

> Three races came together [...] to [determine] the physical and moral genesis of the society of the [*Río de la*] *Plata* [...]: the European or Caucasian as an active part, the indigenous or American as an auxiliary, and the Ethiopian as a complement. From its fusion, this original type resulted, in which European blood has prevailed due to its superiority, constantly regenerating itself by immigration, and on whose side that other mixed race of black and white has grown and improved, which has assimilated the physical qualities and morals of the superior race. (Argentina 1916, I, 198–199, author's translation)

The adjectives 'active', 'auxiliary', and 'complement' implied a hierarchy in which the white element had a superiority that did not derive from its quantitative importance in the total population. More importantly, the 'superior race' not only passed on its physical but also its moral qualities in a discursive logic in which the autonomy of the latter with respect to biological factors was not totally convincing. The metaphorical language of the quote, in particular the image of 'blood', and the polysemy of the term race at the time, however, does not allow clear differentiation of how much of this assimilation derives from biology and how much from social and cultural factors.

Taken as a whole, the census argument/logic related to the ideas of founding fathers of modern Argentina, such as Juan Bautista Alberdi and Domingo Faustino Sarmiento, although with interesting nuances. For Alberdi, author of the famous *Bases y puntos de partida para la organización política de la República Argentina* (1852), the conviction of the existence of a racial hierarchical scale was accompanied by a pro-melting pot policy, founded on the belief that racial mixing would help to dilute the negative traits of the less suitable races. Alberdi's optimism was questioned by various individuals, particularly by the late work by Sarmiento in *Conflictos y Armonías de las Razas en América* (1900). They opposed miscegenation because they considered that the inferiority of blacks and of indigenous peoples would qualitative degenerate the Argentine white race. Though racist connotations within censuses were common to contemporary Western thought, the Argentine census discourse had an integrationist aim that conceived racial mixture as a programmatic and interpretive key to understand the Argentine social process. In that precise sense, the overall interpretation of census-takers moved away from social Darwinism towards a perspective closer to cultural evolutionism.[16]

With the omission of the racial categories in the censuses, in their comments the census-takers emphasised the almost total inexistence of the black population and, in general, of all non-white races:

> The Latin race constitutes [...] the vast majority of the population with 975 per thousand of the total; but the Germanic, Anglo-Saxon, and Scandinavian with the remaining 25, contribute to the improvement of it, giving rise to a new one, by the fusion of its various elements. [...] The Asian race and the black African exist only in small shares, so that their influence is zero in terms of the transformation of the country. The same thing can be said about the Indians. (Argentina 1898, II, XLV–XLVI, author's translation)

The claim that the influence of Scandinavian race was greater than either the black or the indigenous contributions at the end of the century could only be accepted as valid on the hypothesis of a qualitatively different contribution of Scandinavians, based on a classification hierarchical of races, and not in arguments that, like the one cited, erroneously invoked quantitative shares.[17]

The omission of race as a category also led to partial interpretations of important phenomena. Thus, to explain the disappearance of black individuals, the census-takers resorted to explanations focused almost exclusively on mortality. It is not worth dwelling on this point here, which constitutes one of the most successful aspects of George Andrews' (1989) work. It is enough simply to emphasise the multi-causal relationship that existed between the components in the statistical chain: the failure to

[16] On the differences between social Darwinism and cultural evolutionism see Lilia Schwarcz (1993). In the Brazilian case analysed by the author, the predominance of social Darwinism gave rise to a fairly generalised view of *missigenaçao* (miscegenation) as one of the main ills of the population, a pessimistic view that contrasts with the optimism Argentine census-takers had concerning the so-called melting pot.

[17] Despite its underreporting, the share of indigenous peoples in 1914 was 4.7 per thousand inhabitants, while the Scandinavians amounted to 0.6 per thousand.

survey the black population gave rise to an explanation focused on fast-acting factors such as mortality, an argument that justified the omission of the category.

The First Element of Progress

> The foreign population [...] has constituted and constitutes the main force and the first element of progress and labour of the Republic. Alberto B. Martínez (Argentina 1916, I, 201, author's translation)

Parallel to the omission of racial categories, the censuses showed the existence of foreign population, a procedure that began in 1869 and which reached its ultimate expression in 1914. Accounting for foreign groups can be analysed through the criteria in which they were listed (extensive or intensive), the order in which the categories were displayed, and their definition.

Nationality, like other variables that admit many categories, can be presented following an extensive criterion through the exhaustive enumeration of all groups, or through a taxonomy that bundles up the nationalities observed by using one or more criteria. In 1869, the classification followed a geographical criterion: American, European, African, Asian, and no specification. The category 'African' was not a racial one. It meant that the individual was born in the African continent.

The second census used more elaborate forms of data presentation. It alternated between the extensive presentation and the alphabetical order of the nationalities that characterised the 1869 census and the analysis of nationalities bundled up 'by race'. Double registration following an exhaustive enumeration and a grouping of categories also appeared in the tables displaying access to property. While the owners of all nationalities surveyed were registered, information was provided only for the European population.[18] This presentational choice responded to one of the basic features of the censuses of the time: the external use of information intended to serve as propaganda in the migrants' countries of origin. Europeans, whose partially artificial success in accessing property

[18] Argentina (1898, II, tables XVIII y XVII, respectively).

constitutes the *enjeux* of the table, were listed in detail while other immigrants, especially those from bordering country, were not considered.[19] The census' summary information emphasised the same positive aspects of the Argentine social process that were highlighted in the reports produced by the Department of Immigration (*Departamento de Inmigración*) to attract European immigrants.

The third census returned to the unique and exhaustive classification of 1869. The exhaustiveness of 1914 was also manifested in the abandonment of the residual category 'other' that was frequently used in previous censuses. The removal of this category is interesting since, despite including a very low share of migrants, it generally had a greater numerical importance than that of carefully distinguished small groups. Austrians and Portuguese in 1869, and Germans, Austrians, English, and Swiss in 1895, whose presence was less than that of 'other Europeans', appeared carefully listed in the tables without it being possible to determine to which 'others' the residual category referred to. It was clear, however, that the threshold of visibility beyond which a group deserved a place in the statistical table was different for each nationality.

The three censuses recorded the nationalities in strict alphabetical order, except for the Argentine nationality, which always occupied first place. The comparison with other classification systems of the time, such as the complex Japanese system based on the ordering of foreigners according to the existing relations of conflict and alliance with the Japanese state, may be useful to remember that the neutral criterion of alphabetical order was far from universal at the time.[20] Except for the marginal use of the category 'other' and its differential awareness in terms of perceptual thresholds, the alphabetical classification showed the explicit absence, at the level of the national statistical apparatus, of a hierarchy of foreign nations and the agreement with the most modern egalitarian criteria of the Western statistical tradition.

Overlapping with the complete enumeration of nationalities, the census also presented a binary classification that juxtaposed 'Argentines' and 'foreigners'. The results of the binary classification could be presented in

[19] The lack of information on the size of the properties, the concept of 'average ownership', and the lack of consideration of the age of Argentines and foreigners contributed to artificially increasing the proportion of successful foreigners (Otero 2006).

[20] On the logic of the Japanese system, still present in the 1987 census, see Goldstein (1990).

three ways: the combined use of both; the single use of the full classification; and the single use of the binary classification. In the first case no information would be lost, while the second is rare in the census tables. The third possibility involves a significant loss of information since it does not allow further disaggregation. Therefore, it is the most interesting for this research.

The single use of the binary classification within the censuses of 1869 and 1895 occurred when crossing certain variables: marital status, education, occupation, and special conditions such as illness, illegitimacy, cohabitation, religion, and fertility of married women, the latter two only for 1895. The complete classification—alone or combined with the binary one—was applied to the tabulations of the remaining variables. Marital status and occupation were not disaggregated by national group, while access to property was.

The differential use of both classifications shows how the external use of censuses guided the interest in presenting in detail the successful access of immigrants to property. Moreover, the binary classification between Argentines and foreigners regarding occupation, marital status, and fertility induced a homogeneous and non-ethnic image of the labour market and of marital integration, homogenising the process of social reproduction.

The order and quantity of categories are not the only ways to analyse a classification since the definition of the categories occupies an important place. The nationalities defined by the census-takers, for example, followed a political and legal criterion linked to the existence of nation states. Thus, only individuals from countries of an existing state or from a state that was in the final stages of being constituted could be acknowledged as members of a nation. The fact that this form of perception is natural today should not make us forget that it was linked to the construction of nations during the nineteenth century and to the adoption of specific measures taken by political elites.

Indeed, national statistical systems paid special attention to the registration of foreigners. In the Argentine case, the guiding principle was based on the identification between nation and state in a construction in which the existence of the latter legitimised the recognition of the former. For this reason and with few exceptions, for example Germany in 1869, nationalities without the corresponding political structure of the state were not registered as such. Their citizens were registered within the nationality of the state that encompassed them. Obvious legal and technical considerations favoured this procedure, but it would be a mistake to consider that the national statistical apparatus played a passive role in the registration of foreigners.

The statistical system's activity was not limited to the selection of categories. It appeared in the first component of the statistical chain: the one that links enumerators and those being registered. The Passenger Entry and Exit files for the period 1841–1860, years before the first population census, indicated that "care should be taken when Basques (sic) appear, to see those who are Spanish Basques or French Basques, and to write down the Spanish as Spanish, and the French as French."[21] Likewise, Fernando Devoto (1994, 135) shows how the uniformity of the records of the lists of the 1855 City of Buenos Aires census allows us to infer that the nationality of the immigrants was externally assigned by the census-takers. In this way, the identity registered by the enumerator replaced individual's self-identification.

National censuses operated in the same way. According to the historiography on Argentine immigration, identification with the nation state was far from being the rule of the inhabitants of nineteenth-century Europe. Moreover, in a significant share of cases, identity was defined by belonging to regional units of greater cultural homogeneity and community of interests rather than to the elephant-like states in which they were embedded. For this reason, even in 1914, enumerators complained that foreigners "in most cases only recorded the place of birth by town and even by village" and not by nation, which made it necessary to resort to "a geographical dictionary to establish the nation they belonged to" (Argentina 1916, I, 52, author's translation). Through these mechanisms of external identification and transcription of responses to a national grid, official statistics proceeded to identify French, Italian, or Spanish while creating Argentines. As in the United States, an unanticipated effect of these operations was that the question of national origin favoured many immigrants to identify with their countries of origin, which united ethnic groups. Having learned in the new country that they belonged to a nation, many immigrants immersed their provincialisms in a broader patriotism and their local dialects in a language (Petersen 1987, 218–219).

21 "Legajo de Entrada y Salidas de Pasajeros, 1841–1860," Archivo General de la Nación (author's translation).

The legal view that led to the definition of identities, even in opposition to the self-identification of individuals, was also applied to the Argentine population with important consequences. The adoption of the *jus solis*, which determined that the nationality of an individual is that of the territory in which they were born and not that transmitted by their parents (*jus sanguinis*), had its counterpart in defining and enumerating the Argentine natives. As Francisco Latzina pointed out in an apparently paradoxical phrase "there are no foreign children born in this country" (Argentina 1916, IV, 526, author's translation). Within this legal framework, the children of foreigners or of mixed unions became Argentines both for the law and for the census from the first generation. This ruled out alternative forms of registration such as in the Canadian and US systems that distinguished the population born in the country according to their migratory origins.[22]

This country comparison illustrates the autonomy of statistical systems to promote 'local' forms of measurement, regardless of their similarities in other levels. Even though all three were immigration countries that logically adhered to nationality based on *jus solis*, they set up inquiry mechanisms of very different historical depth. In Canada, measurement focused on national origins, whatever the generational distance with the first immigrants, while in the United States the measurement was up to the second generation, and in Argentina it was limited to immigrants. The same can be observed in other ethnic indicators, such as the spoken language, which ranged from a sophisticated inquiry system in Canada to an absolute absence of questions on the matter in the Argentine case.[23] A similar situation occurred with religion, with the

[22] Referring to the 1910 US census, Alberto B. Martínez stated that "the question about the colour or race of the inhabitants, and the place of birth of the father and mother of the registered person in the United States census are not important to us because, happily, we do not experience either a worrying racial question nor are our ethnic roots still so deep that they impose on us the need to investigate the origin of foreigners" (Argentina 1916, I, 41, author's translation).

[23] The question on language is not secondary. As the National Indigenous Census illustrates, a significant share of indigenous peoples from the provinces of Chaco and Misiones did not speak Spanish as late as 1965. The same conclusion can be reached if the regions that practice bilingualism are considered; for example, the wide area where *guaraní* was spoken.

aforementioned exception of the 1895 census. Thus, maximum ethnicisation in Canada, intermediate in the United States, and zero in Argentina constituted three levels of statistical ethnicisation of the population.[24]

Consequently, the transformation of foreigners into Argentines was forcibly operated in the span of a single generation, through a process of homogenisation of children born in the country, regardless of their national origin and their distance from the migrant ancestors. For this reason, the integration and assimilation of Europeans through "family alliances" (much slower according to recent research) would be "almost complete from the first years of their arrival" (Argentina 1898, II, XLIV, author's translation). The *a priori* nature of this statement is evident in its inferential nature since the census data, with the sole exception of the unbalanced sex ratio of migrants, were not relevant to warrant such a conclusion, especially regarding the speed that characterised the Argentine melting pot.

Conclusion

The joint analysis of Indians, blacks, and immigrants shows how the population was conceived during the formation of the Argentine statistical system and provides an explanation of its long-term symbolic effects. Two of these effects are the homogenisation of a heterogeneous population and the construction of a certain image of the Argentine nation. In both cases, population censuses were not and could not have been—given their character of intellectual construct—a passive measurement instrument, since the selection and definition of the categories of analysis helped create an image of the nation that would have to be partially autonomous from the historical process. This autonomy also implied a connection—ambiguous and partial but never absent—with the available empirical data, which distanced the census reconstruction from the most radical interpretations of the concepts of 'invention' or 'imagination'. Indeed, even considering the ideological premises analysed, the nation of the census-takers was not an arbitrary construct but a display of the

[24] For an analysis of the US American censuses and the difficulties of measuring ethnicity see William Petersen (1987). The Canadian case is analysed by Jean-Pierre Beaud and Jean-Guy Prevost (1996) and Bruce Curtis (2001).

features that they considered essential in Argentine socio-demographic development.[25]

The censuses' homogenising effect had its origins in a legal classification that differentiated individuals according to their belonging to national entities identified with independent states. The basic result of this procedure was twofold. On the one hand, it helped standardise the discourse of heterogeneous groups of foreign population through the external attribution of national identities and the blocking of alternative forms of self-identification, whether regional as in the Italian case or inter-states as in the Basque case. On the other hand, it made the native population uniform thanks to a series of political and conceptual operations. These operations included denying that indigenous peoples were pre-existing nations, a recognition that would have been contradictory with the hegemonic territorial claims of the state. These operations also implied the abolition of categories of colonial origin such as race as well as the homogenisation of the generations of immigrants born in the country through *jus solis* and the non-inclusion of generational markers, as in Canada and the United States. In this perspective, the mechanisms of standardisation of Argentines and foreigners mutually reinforced each other as both were based on a political logic that identified the state with the nation and vice versa.

The choice of a census grid that ruled out accounting for certain population groups did not obey numerical visibility criteria, although this reason was sometimes invoked, since the black and especially the indigenous population were quantitatively superior to many of the European groups. On the contrary, the adoption of this grid was based on the explicit recognition of the enormous technical difficulties in measuring the population according to racial criteria. In this sense, and despite the racist nature of many premises of the census discourse, the decision based on technical minimalism implied the conviction that the distinction of people according to racial criteria visible 'from the outside' was nothing short of chimerical. The different aspects of the census discourse implied

[25] The shares of people who self-identified as indigenous and Afro-descendant in the 2010 Argentine census were 2.4 and 0.4%, respectively, much lower than those of other countries in the region. On the reincorporation of these categories from the 2001 census, see Luis Angosto Ferrández et al. (2012) and, above all, Loveman (2014), who proposes a long-term view for Latin America as a whole, while also paying attention to the influence of external factors, which are more difficult to perceive in the case studies.

a simultaneous process of racism and cultural evolutionism in the general theoretical and interpretative sphere, and of denial of the possibility of differentiating individuals by physical appearance or self-definition in the empirical level.

Differentiating inhabitants based on their nationality allowed the construction of an egalitarian and objective census grid based on the alphabetical order of categories and a progressively exhaustive enumeration of groups, for example. The omission of ethnic and racial categories of the colonial past contributed, in the same sense, to an in principle non-discriminatory presentation of the populations at stake, thanks to a liberal-political statistical paradigm that, contrary to the colonial conception, only recognised individuals as the basis of society.

According to the genealogical matrix that supports it, the census interpretation can be characterised as a self-celebratory speech that sought to present itself to the rest of the world by highlighting those specificities that characterised the Argentine nation in relation to other countries in the region: a country of natives and of foreigners who were integrating rapidly. However, the modern and egalitarian characteristics of the national grid were contradictory to the more general ideological considerations of census-takers, such as their assessment of non-white races and the problems posed by their integration and further reduction. As has been said, enumerators adhered to that school of thought that, guided by the principles of natural selection and trusting in the regenerative superiority of the white race, fervently adhered to racial integration, an attachment that implied a departure from the more reactionary and unscientific concepts of social Darwinism.

The underlying contradiction between a liberal reading of egalitarian character and a hierarchical racial view, however, would not pose major interpretive problems for the census-takers, thanks to the dissimilar proportion of populations at stake and the positive and particularly rapid effects attributed to mortality, miscegenation, and the melting pot, the master key of the census paradigm. The omission of racial categories, the non-distinction between generations of immigrants, and the dichotomy between Argentines and foreigners in key variables of the social reproduction process like occupation, marital status, and fertility reinforced this interpretation by liquefying the data necessary for its eventual authentication. It is precisely thanks to this dissolution, rather than to the results of the process, that the political dimension of the censuses reveals its deepest meaning.

Ultimately, the combined effects of legal classification and the extraordinary speed of the integration of races and nationalities allowed the censuses to operate, always on a symbolic level, on the transformation of a plural and diverse mass into a relatively homogeneous population. Ideological discourse and forms of measurement were mutually implicated to produce that 'homage to the nation' that are the population censuses. In such a way, these fantastic works were not limited to measuring the present and the still fresh traces of the recent past, but at the same time contributed to establishing a genealogy through which the statistical system projected out of the country and into the future a determined image of the Argentine nation.

References

Aguirre Beltrán, Gonzalo. 1970. "The Integration of the Negro into the National Society of Mexico." In *Race and Class in Latin America*, edited by Magnus Mörner. Nueva York: Columbia University Press.

Anderson, Benedict. 1983. *Imagined Communities. Reflections on the Origin and Spread of Nationalism*. London and New York: Verso.

Anderson, Margo J. 1988. *The American Census. A Social History*. New Haven: Yale University Press.

Andrews, George Reid. 1989. *Los afroargentinos de Buenos Aires*. Buenos Aires: Ed. de la Flor.

Angosto Ferrández, Luis Fernando, and Sabine Kradolfer, eds. 2012. *Everlasting Countdowns: Race, Ethnicity and National Censuses in Latin American States*. Newcastle upon Tyne: Cambridge Scholars Publishing.

Appadurai, Arjun. 1996. *Modernity at Large*. Minnesota: University of Minnesota Press.

Argentina. 1872. *Primer Censo de la República Argentina, verificado los días 15, 16 y 17 de setiembre de 1869*. Buenos Aires: Imprenta del Porvenir.

Argentina. 1898. *Segundo Censo de la República Argentina, mayo 10 de 1895*. Buenos Aires: Taller Tipográfico de la Penitenciaria Nacional.

Argentina. 1916. *Tercer Censo Nacional, levantado el 1° de junio de 1914*. Buenos Aires: Talleres Gráficos de L.J. Rosso y Cía.

Beaud, Jean-Pierre, and Jean-Guy Prévost. 1996. "Immigration, Eugenics and Statistics: Measuring Racial Origins in Canada (1921–1941)." *Canadian Ethnic Studies* 28 (2): 1–24. https://search.proquest.com/docview/215640280?pq-origsite=gscholar&fromopenview=true.

Bertoni, Lilia Ana. 2001. *Patriotas, cosmopolitas y nacionalistas. La construcción de la nacionalidad argentina a fines del siglo XIX*. Buenos Aires: Fondo de Cultura Económica.

Biagini, Hugo. 1995. *La Generación del 80*. Buenos Aires: Losada.

Curtis, Bruce. 2001. *The Politics of Population: State Formation, Statistics, and the Census of Canada, 1840–1875*. Toronto: University of Toronto Press. https://doi.org/10.3138/9781442682108.

Daniel, Claudia. 2011. *El observatorio social. Estado, censos y estadísticas oficiales en la Argentina (1869–1914)*. Saarbrücken: Editorial Académica Española.

Devoto, Fernando. 1994. *Le migrazione italiane in Argentina. Un saggio interpretativo*. Napoli: L'Officina Tipografica.

Germani, Gino. 1962. *Política y sociedad en una época de transición. De la sociedad tradicional a la sociedad de masas*. Buenos Aires: Paidós.

Goldstein, Joshua. 1990. *Au dedans du dehors: Une reconstruction typologique de l'étranger vu du Japón*. Paris: Editions EHESS.

González Bollo, Hernán. 2014 *La fábrica de las cifras oficiales del Estado argentino (1869–1947)*. Bernal: Universidad Nacional de Quilmes Editorial.

Hale, Charles. 1991. "Ideas políticas y sociales en América Latina, 1870–1930." In *Historia de América Latina*, edited by Leslie Bethell. Barcelona: Crítica.

Halperín Donghi, Tulio. 1980. *Proyecto y construcción de una nación. Argentina, 1846–1880*. Caracas: Biblioteca Ayacucho.

Le Bras, Hervé. 1988. "La Statistique Générale de la France." In *Les lieux de la mémoire*, edited by Pierre Nora. Paris: Gallimard.

Loveman, Mara. 2014. *National Colors. Racial Classification and the State in Latin America*. New York: Oxford University Press. https://doi.org/10.1093/acprof:oso/9780199337354.001.0001.

Marucco, Dora. 1996. *L'amministrazione della statistica nell'Italia unita*. Roma-Bari: Editori Laterza.

Merllié, Dominique. 1989. "La construction statistique." In *Initiation à la pratique sociologique*, edited by Patrick Champagne, Rémi Lenoir, Dominique Merllié, Louis Pinto. Paris: Dunod.

Nacach, Gabriela. 2012. "El *enigma* de Nuestra Señora de la Candelaria. Tierra del Fuego *libre de indios* en el relevamiento censal de 1895." *Atek Na* 2: 121–164. https://plarci.org/index.php/atekna/article/view/124. Accessed 23 March 2021.

Otero, Hernán. 2006. *Estadística y Nación. Una historia conceptual del pensamiento censal de la Argentina moderna, 1869–1914*. Buenos Aires: Prometeo.

———. 2018. "Socio-History of Statistics on Latin America: A Review." *Histoire & Mesure* 33 (2): 13–32. https://doi.org/10.4000/histoiremesure.7934.

Patriarca, Silvana. 1996. *Numbers and Nationhood. Writing Statistics in Nineteenth Century Italy*. Cambridge: Cambridge University Press. https://doi.org/10.1017/CBO9780511523458.

Petersen, William. 1987. "Politics and the Measurement of Ethnicity." In *The Politics of Numbers*, edited by William Alonso and Paul Starr. New York: Russell Sage Foundation.

Sánchez Albornoz, Nicolás. 1994. *La población de América Latina, desde los tiempos precolombinos al año 2000.* Madrid: Editorial Alianza.

Schor, Paul. 2009. *Compter et classer. Histoire des recensements américains.* Paris: Editions EHESS.

Schwarcz, Lilia Moritz. 1993. *O espetáculo das Raças. Cientistas, Instituições e Questào Racial no Brasil, 1870–1930.* Sao Paulo: Companhia das Letras.

Terán, Oscar. 1987. *Positivismo y Nación en la Argentina.* Buenos Aires: Punto Sur.

CHAPTER 6

Loose Numbers: Political Centralisation and Statistical Fragmentation in Colombia, 1886–1930

Victoria Estrada Orrego

This chapter explores the production of official statistics in Colombia between 1886 and 1930, a period conventionally known as the *República Conservadora* (Conservative Republic). As in other Latin American countries during the nineteenth century, local conservative forces promoted a series of reforms aimed to centralise political power. This was their way of counteracting liberals, but also a decision that expressed their vision of power and the physiognomy of the state. By means of the new constitution of 1886 Colombian conservatives reversed the federalist regime liberals had established in 1863, limited the influence of the National Congress, and restricted individual freedoms, as the Catholic Church recovered its traditional place within society. The production

V. Estrada Orrego (✉)
Instituto Tecnológico Metropolitano, Medellin, Colombia
e-mail: victoriaestrada@itm.edu.co

C. T. Lanata-Briones et al. (eds.), *Socio-political Histories of Latin American Statistics*, Studies of the Americas,
https://doi.org/10.1007/978-3-030-87714-9_6

of official numbers was also part of this centralising agenda. Statistics, however, proved to be a practice difficult to tame. During these years, the agency in charge of collecting numbers was systematically part of different ministries, changed its name several times, and was headed by more than ten different people. Several laws were issued, and there were continuous efforts to incorporate local authorities into a national and standardised network of knowledge production. None of these measures, however, were enough to cope with the geographic, bureaucratic, and political obstacles the Colombian state had to face in trying to forge its hegemony across the territory.

Conservative authorities were not alone in the attempts at centralising the production of official statistics. The increasing internationalisation of the field at the turn of the century was another factor pushing for this transformation. The Colombian state was not yet ready to assimilate the conceptual tools and the bureaucratic routines the global statistical system recommended. However, these guidelines were crucial in promoting the normalisation of Colombian official numbers. The discourse of local physicians concerned with the scarcity of knowledge regarding diseases, causes of death, and related sanitary occurrences played a similar role. Once again, the signing of international agreements, and the development of a medical discourse that showed that living conditions were not a matter of fate or charity, but of public interest, paved the way to the conviction of the necessity of modernising statistics, which in turn reinforced the idea that such modernisation depended on centralising its production.

What we know about the history of Colombian official statistics is mostly written by former statistics officials interested in describing the legislation and the trajectory of the main institutions involved in these activities, which usually was a way of accounting for their own experience as civil servants (Rodríguez 1928; Otero D'Costa and Arrubla 1935; Lleras Restrepo 1938; Hermberg 1941; Charry 1954; Vidales 1978). Given that focus and also the institutional milieu in which most of such reports were produced, these works tend to be more narrative than interpretive. As such, they narrow the scope of the production of statistics to the mere accumulation of norms, the succession of directors, and the listing of bureaucratic achievements.

This chapter offers an approach to the history of the Colombian statistical apparatus looking at norms, directors, and institutional achievements, but always putting them in connection with determinants coming from the political, intellectual, and social fields. In doing so, it covers the years of the *República Conservadora* analysing the interaction between the physiognomy of the territorial administrative system and the production of official statistics, in particular how that interaction gave form to the always contingent relationship among practices, norms, authorities, and institutions (Beaud and Prévost 2010).

Although the Colombian state produced a wide range of numbers, this chapter follows closely the transformations on demography and health statistics. The trajectory of both fields offers an interesting case of institutional paradox: an increasing divergence under centralising efforts. The ideological affinity between conservative elites and the Catholic Church somehow reinforced such paradox. As historical collectors of information on births, marriages, and deaths, parish priests continued playing a critical role in the history of vital statistics. Given the limitations of the state bureaucracy, and the weakness of its personnel to claim the monopoly over such data, conservatives relied heavily upon the informational structure of the Church. In the short term, this choice allowed the state to keep a rhythm and an order in the gathering of social facts, but in the long term it delayed the configuration of a state apparatus able to perform these tasks under a secular logic.

Demography and health statistics are thus privileged arenas to collect clues on the obstacles the authorities faced in centralising the production of numbers due to the combination of geographic constraints, the scarcity of trained cadres, population distrust of statistics, political violence, bureaucratic overlapping, and financial problems. With the aim to foster the development of new, less self-congratulatory approaches to the history of Colombian statistics, this contribution dialogues with some recent works devoted to the Colombian censuses conducted in the first half of the twentieth century, in particular the 1912 population census (Estrada 2017; Prieto 2005), and with those specialised studies on the history of diseases like leprosy and tuberculosis (Obregón 2002; Martínez and Guatibonza 2005; Estrada et al. 2016).

This chapter is divided into three parts. The first one studies the transformation of Colombian statistical agencies in the first two decades of the conservative rule, tracing its impact on demographic and health statistics. The second part analyses the process of administrative centralisation and the efforts to dovetail such institutional move with the production of official statistics according to global standards. The third section explores how the continuous reorganisation of the Colombian statistical field ended up fostering the fragmentation of efforts and the institutional divergence between population and health statistics.

The Production of Official Statistics in the First Years of the *República Conservadora*

Between Physicians and Parish Priests

Throughout the nineteenth century, Colombian society experienced numerous civil wars which led to endless administrative reorganisations. As part of these reforms, several laws and decrees regarding the production of statistics were issued, most of them to determine the organisation of population censuses and to establish the electoral roll. Voting rights were attached to income, property, and literacy requirements. Hence, statistics became crucial for the electoral process, as it decided the allocation of seats in the Senate and the Chamber of Representatives (Vidales 1978, 58).

During the years of liberal hegemony (1863–1886), when the country was organised as a federal republic, statistical activities were located first in the Ministry of Government and Foreign Affairs, and later in the Ministry of Development. The state was able to produce figures on foreign and maritime trade and even organise a population census in 1870, the sixth and last one of the century, given that the previous ones had been conducted in 1825, 1835, 1843, 1851, and 1864. However, these were years of instability in statistical matters. The directors of the specialised agency were replaced almost every year, and continuous budget cuts forced the suppression of administrative post that affected the regularity of the statistical inquiries (Charry 1954, 31–40).

After winning the 1886 election, the Conservative Party established the first institution formally devoted to health matters at a national level, the Central Council of Hygiene (*Junta Central de Higiene*). The council was divided into four commissions, one of which was responsible for preventing and combating endemic infections and epidemics, as well as for studying mortality and its causes, and organising the production of medical statistics (Junta Central de Higiene 1887). This office was located under the supervision of the Ministry of Development.

Ever since the International Sanitary Conference held in Washington in 1881, Colombian physicians stressed the need to establish a centralised agency in charge of collecting and publishing figures on prevalent illnesses, rates of mortality, and public health matters. The situation was critical and unsettling: out of the 33 attendants to the conference, Colombia was the only country without such an office (Márquez 2005, 13–14). That is why the creation of the Council of Hygiene was a significant event. One of its first measures was to establish protocols of sanitary supervision at Colombian ports and forcing quarantines for individuals that suffered pestilential diseases such as the bubonic plague, Asian cholera, and yellow fever. This creation was not only a sanitary decision, but also a commercial one since its existence was mandatory for taking part in international trade. Given the expansion of the coffee-exporting economy, the council played a critical role in the modernisation of the sector (Obregón 2002, 163).

In parallel to the organisation of the Central Council of Hygiene, in 1887 conservatives signed a Concordat with the Holy See allowing the use of parochial certificates as official proof of citizens' civil status (República de Colombia and Gobierno Nacional 1887). This was the first step of a series of measures that ended up placing the entire set of vital statistics under the control of the Catholic Church, reassuring the authority of parish priests as official administrators of births, marriages, and deaths records. In the absence of a civil bureau, the state assumed its dependence on religious bureaucracy, which had undeniable advantages *vis-à-vis* its own networks. The final episode of this series of agreements established that parish priests had to send reports to the civil authorities every six months (República de Colombia and Gobierno Nacional 1892). Even though it was not a secular solution, it paved the way to a form of centralisation.

First Reform: The Statistics Office

In what seems to be a clear symptom of the lack of coordination in the production of official statistics, in 1888 Conservatives also decided to create the Statistics Office (*Oficina de Estadística*). The Colombian Pedro Salcedo Ramón was its first director. In practical terms, this creation allowed the state to gather a small cohort of specialised employees who were scattered in different administrative units and transfer them to the Ministry of Development, where the new agency was located, along with the Central Council of Hygiene (República de Colombia and Consejo Nacional Legislativo 1888). An overview of the scope of the topics assigned to the Statistics Office allows us to understand its objectives and also its place in relation to the Council of Hygiene and the tasks entrusted to the Catholic Church. The office oversaw the collection of data on railroads and roads, patents and patent licensing, public works, lighting and cleaning of public streets, maritime and inland waterway shipping, agricultural and industrial activities, among others (República de Colombia 1889).

The creation of the Statistics Office was thought as a way of concentrating efforts and moving towards a better coordination in the measuring of economic growth. However, the first institutional design was not able to take off and two years later, in 1890, the office was reformed as part of a major reorganisation of the ministry. Despite the problems initially detected, such as shortages of resources and personnel, the Statistics Office, now turned into a ministry division, was entrusted with new duties. These tasks included the collection of information on commerce, industry, maritime activity, natural exploitable resources, and road networks, as well as data on population, religion, public education, land registry, debts, mortgages, banks, charities, crime, and penitentiary services (República de Colombia and Ministerio de Fomento 1891). Albeit this list of items coincided with the index of topics usually covered in the statistical yearbooks at the time, for the local bureaucracy this horizon was too ambitious. The outcome was somehow predictable. Towards 1892, trade statistics was the only topic with regular publications.

That outcome, of course, has an explanation, and it has to do with the trajectory of the head of the Statistics Office in this initial period: Anthony John Schlesinger (1835–1897). Born in London to a German father and English mother, Schlesinger arrived in Colombia as a representative of a transatlantic insurance company based in Berlin. As many other foreigners in the region, he made his way into the national statistical service covering the lack of qualified personnel. Even though his appointment can be explained from his friendship with influential people, he was one of the foreigners that played a significant role in the development of the seminal institutions of the Colombian statistical system (Estrada 2015, 26). What was possible (and not) in this period of organisation depended closely upon the training, interests, and capacities of those entrusted with the mission of establishing order and regularity within a still fragmented field.

It is worth noting that the Statistics Office and the Central Council of Hygiene did not worked jointly in these years, despite the fact that both entities were part of the same unit, the Ministry of Development. In an early sign of a dividing line that would become critical in the following years, the figures produced by the Statistics Office were mainly attached to commercial and financial affairs, far from the demographic leaning this type of agencies had elsewhere. A revealing fact can speak for this assertion. Even though conservative authorities tried to foster the centralisation and modernisation of the statistical regimen, no population census could be conducted until 1905, more than a decade after this ambitious reform.

Second Reform: The Central Statistics Office

Given the scarce results of the previous reorganisation, in 1892, the statistics section of the Ministry of Development was transformed into to the Central Statistics Office (*Oficina Central de Estadística*). This new denomination did not mean a radical change in terms of its research agenda, not at least when considering the scope of the matters that were part of the tasks of the previous institution (República de Colombia and Ministerio de Fomento 1893). However, it was a new attempt to reach the centralisation that the conservative governments aimed for.

Once again, the agency was put in the hands of a foreigner: Sophus Höeg Warming. Of Prussian descent, Warming arrived in Colombia in 1891 to train cadets at the Military School (*Escuela Militar*). In tune with conservative aspirations, the Colombian Military School sought to create a central army aimed at monopolising the use of force. This was an idea liberals and federalists had always opposed to, but which, in a centralising context, gained momentum. The Military School began operating in March 1891, but students' poor performance and severe financial problems prompted its closure in late 1892. Warming, who was already connected with the Statistics Office, seemed a perfect candidate to lead the new institution (Atehortúa and Vélez 1994, 36–38; Rey 2008, 159).

Inspired by Prussia's statistical model, Warming aimed to establish a well-trained agency to improve national statistics. The Prussian model was known for establishing hierarchical and centralised data registry structures that assured technical precision and central coordination (Desrosières 1993, 219–224). The primary aim of the Central Statistics Office was to "compile all essential data" in order to portrait more precisely "the population, wealth, civilisation, and power of the Republic." The office had six employees, one of them in charge of preparing and publishing foreign trade statistics, which was always considered a crucial topic. The capital city of each department—the political/administrative units of Colombia—had to maintain a local agency devoted to gathering and sending information to the office's headquarters. In turn, with all the data coming from the territorial units, the Central Statistics Office had to prepare the Quarterly Bulletin of the National Statistics of Colombia (*Boletín trimestral de la estadística nacional de Colombia*). Among their duties, town and city mayors oversaw the provision of monthly population data, which they had to register on the forms timely established by the Central Statistics Office. However, given the country's poor road network, and the difficulties in printing the forms, the information not always arrived on time, if it arrived at all. Usually, the civil servants of the statistical office received incomplete forms, or forms with data that did not overlap with what had been requested (Warming 1892, x–xx).

Despite these obstacles, four issues of the Quarterly Bulletin were published in 1893 and one more in 1894. The content of the bulletins was quite varied, ranging from import and export figures and information on mining exploitation to marriage and mortality monthly statistics across the country. But the Bulletin was also a place in which the staff of the Central Statistics Office made references to the pitfalls and shortcomings in the collection of reliable data. There were constant complaints about forms incorrectly filled or containing false information, all with the sole purpose—according to Warming—of dodging the penalties established by law for careless bureaucrats (*Boletín trimestral de la estadística nacional de Colombia*, 1–5). In sum, because of the deficiencies in the performance of departmental civil servants, derived from their ignorance or lack of interests, the scientific aspirations of the Central Statistics Office were often not meet.

Although the collected information was always incomplete and irregular, it still allows us to decipher the variables and categories chosen by the Central Statistics Office to record information. Statistics on marriage, for instance, were organised by sex, age range, skin colour (white/coloured), religion (Catholic, dissident), 'physical malformations,' literacy level, kinship relations between spouses, and number of legitimate children per marriage. The binomial Catholic/dissident, the question on kinship bonds, and the quantification of legitimate children were certainly influenced by the close relationship with the Catholic Church. The question on 'physical malformations,' in turn, derived from the idea that such conditions could have consequences for the entire country, determining people's ability to work efficiently, or in the transmission of hereditary diseases to future generations, which was crucial for a nation eager to have a strong population (Warming 1893, 3–36).

The Quarterly Bulletin also included data on causes and places of death and medical assistance before death, being probably the only nineteenth-century source of this sort covering several departments. Regarding places of death, it showed that at the time more than 90% of people passed away outside a hospital. Warming explained this as the consequence of a hospital deficit, combined with the fact that most people avoided them because they were seen as places to die. The fact that most deaths occurred outside hospitals certainly influenced the records of causes of death, which did not usually follow a nosologic scheme, but an empirical criterion. Moreover, parish priests failed to report on these affairs or reported them irregularly. Warming systematically insisted on the need to have physicians recording this data and urged mayors to gather and send it to the office's headquarters. Accumulating knowledge on epidemic diseases, particularly of the most frequent ones, was, according to Warming, one of the means "to avoid [...] or cure them" (Warming 1893, 39–55, author's translation).

Due to unknown reasons, Warming's term ended in 1894. Likewise, the publication of the Quarterly Bulletin was discontinued due to budgetary problems. The Colombian congress passed a new law in December 1896 reorganising the national statistical service, probably as a response to the abrupt interruption of Warming's administration and probably to solve some of the problems he had detected. In any case, this was a new attempt at creating a national network of data collection, always following a centralist logic. One of the key measures of the 151 Law was the creation of statistics offices in each department and the establishment of permanent statistics commissions in each municipality under the direction of the corresponding mayor. In turn, the entire statistical service was reallocated to the Ministry of Government and Enrique Arboleda Cortés, a man of arms, was named as its new director.

A (Failed) Demographic Turn

Fostering the production of reliable data on deaths and its causes was one of Enrique Arboleda Cortés' concerns. In May 1899, he sent a letter to the Dean of the School of Medicine and Natural Sciences of Bogotá, requesting his opinion on the Bertillon classification system. This system had been proposed by Jacques Bertillon, director of the Municipal Statistical Office of Paris (*Bureau Municipal de Statistique de Paris*) to categorise diseases and standardise their names. The proposal was first discussed at the Chicago Conference of the International Statistical Institute in 1893 and adopted by the American Public Health Association a few years later. Adopting this system was crucial, according to Arboleda Cortés, not only to rightly register and study diseases, but also to refine cause-of-death statistics. He also asked for advice on how to set it in motion considering the shortage of physicians in Colombian municipalities. This would mean, of course, a new bureaucratic challenge, because governments at the city level had to organise hygiene councils to attend to the scarcity of medical personnel. The positive response from the Dean of the Medical School was based on the usefulness of the system, but also on the necessity of attuning the production of local statistics to the criteria and the categories in use overseas (Arboleda Cortés 1899; Castañeda 1899, 793).

The adoption of this classification system faced several setbacks. As expected, forms were not filled out properly in rural areas, where medical affairs were usually controlled by amateur physicians or locals with no scientific training. It was easier in major cities, where professional physicians helped actively in promoting the use of the new nomenclature. Arboleda Cortés remained optimistic about the possibility of introducing the use of the Bertillon system, although the first experiences were not successful. In his opinion, errors were an "irremediable ill" since the "popularisation of science" depended both "on time and education" (Castañeda 1899, 793, author's translation). Even though the Central Statistics Office kept asking municipal hygiene councils to promote the new policy and enforce its assimilation, local administrations proved to be highly inefficient in the production of quantitative data. There were gross disparities among municipalities and departments, between urban and rural areas, and within regions with and without professional physicians.

The outbreak of the Thousand Days' War (1899–1902)—by means of which the most radical cadres of the liberals tried to regain power or at least retain their electoral rights before a new onslaught of the opponents—stopped the works of the emerging statistical system almost completely. The economic and social effects of what ended up being a long and fierce conflict fractured the realm of politics and the most basic bureaucratic routines (Melo 2020, 173–179). However, this did not prevent the Central Statistics Office from publishing in 1902 some statistics on commercial activities for the period 1897–1898. Albeit there was no time nor energies to do research on demography during these years, trade was always understood as an unavoidable subject. These conditions did not change substantially towards 1905, when a new population census was carried out, the first one in more than three decades. Several setbacks, derived from the combination of bureaucratic fragmentation and lack of experience, affected the quality of the results, which were severely criticised by the authorities. Symptomatically, the corresponding report was published more than a decade later (Estrada 2017, 143).

The 1905 population census also marked the end of Arboleda Cortés' term as head of the Central Statistics Office. He was replaced by Vicente Larra. In the following year, a new reform was implemented, seeking once again to merge in a single office the statistics and statisticians scattered throughout the state bureaucracy. In turn, the entire statistical system was put under the Ministry of Finance and the Central Statistics Office was renamed as the General Board of Statistics (*Dirección General de Estadística*). Larra oversaw the editing of a volume with some of the figures collected by his predecessor. Published in 1907, the book devoted most of its pages to statistics on foreign trade. Data on births and mortality was scarce, restricted to Bogotá and three out of the nineteenth departments in which the country was divided. Larra's diagnosis was clear: there would be no reliable data if the system continued to depend on parish priests, who did not show interest in collecting and sending complete and useful records to mayors and departmental governors. Without an autonomous civil agency, the production of statistics would remain a far horizon for the government (República de Colombia and Dirección General de Estadística 1907, 59–68).

The transformations in the physiognomy of the statistical system during the first two decades of the *República Conservadora* were permanent. In a decade and a half, the main statistical agency had five directors and was under the supervision of three different ministries, Development, Government, and Finance, as Table 6.1 shows. These changes are clear indicators of the difficulties the national government faced in enforcing routines and norms that would allow the consolidation of an efficient bureaucracy. Likewise, the Thousand Days' War also affected the incipient production and circulation of statistics across the country, particularly in relation to demography. As it was usual elsewhere, trade statistics tended to follow a different, more regular path. This was good

Table 6.1 Institutional changes in the Colombian statistical system (1888–1923)

Period	*Name of the statistical agency*	*Location*	*Directors*
1888–1892	Statistics Office	Ministry of Development	Pedro Salcedo Ramón (1888) Anthony John Schlesinger (1891–1892)
1892–1895	Central Statistics Office	Ministry of Development	Sophus Höeg Warming (1893–1894)
1896–1905	Central Statistics Office	Ministry of Government	Enrique Arboleda Cortés (1896–1905)
1906–1923	General Board of Statistics	Ministry of Finance	Vicente Larra R (1905–1909)

Source Author's elaboration

news because these figures placed Colombia in the international arena. Regarding population and living conditions, however, ignorance was the norm.

Scenes of Administrative Centralisation

The General Board of Statistics

Seven years after the 1905 population census, President Carlos E. Restrepo (1910–1914) ordered the design and implementation of a new national enquiry. This was good news, but in the short term it meant little for the systematic production of demographic and health statistics, which remained marginal topics for public administration. Physicians insisted on the importance of this knowledge and fiercely criticised the government for not fostering its development. These pressures not only came from civil servants and professional physicians, but also from representatives of the external trade sector. Indeed, every major nation that relied heavily on foreign trade needed to produce reliable statistics, since the flow of economic, industrial, and human capital depended on them. Not having up-to-date figures only delayed the establishment of trade links with the most advanced nations (Troconis 1912, 7).

Although Colombia had a specialised agency devoted to producing statistics on illness and death, the Central Council of Hygiene, and had joined the International Sanitary Conferences, there were still serious gaps in this data. The outbreak of the First World War did not help to improve the situation. Restrepo's successor, José Vicente Concha, had to manage the deterioration of the terms of trade and the reduction of the fiscal power. As the need for new sources of financial increased, the government implemented several tax reforms implied changes in the General Board of Statistics (Arizmendi 1989).

Two laws passed in 1914 and 1916 were a new attempt to centralise the direction and the services provided by the General Board of Statistics. These legal tools made possible for the board to force individuals and public entities to provide the data it needed for its proper functioning. Both laws aimed to overcome, once again, the increasing gaps in the quality of the information each department provided. Albeit the most serious cases were those units which did not send reports at all, there were still problems with those that did send data, given that the information they provided was usually rudimentary, not standardised, difficult to compare, and even to aggregate (República de Colombia and Dirección General de Estadística 1917, VII).

The Anuario Estadístico

One concrete result of this new attempt at statistical centralisation was the *Anuario Estadístico 1915* published in 1917. This statistical yearbook included data on population, private and public education, rail and river transport, sanitary inspections of ports and ships, banking activities, public services, public finances, mail, and telegraphs. It also recorded information concerning salt mines, crops, livestock, slaughterhouses, land registry, notaries, judicial territorial divisions, weather, demographics, healthcare, and religion. In what was a positive innovation, cause-of-death statistics were documented according to the Bertillon nomenclature. Data collected during the 1912 census allowed the agency to produce some population tables that were crucial for making comparisons within the country and with other nations (República de Colombia and Dirección General de Estadística 1917). Given the inexistence of national demographic statistics prior to 1915, which made impossible to measure changes over time, director Alberto Schlesinger took international comparisons as a useful exercise not only to understand Colombia's population in the world state, but to place it as part of the world. This operation was also performed using data on the primary causes of death in other countries (República de Colombia and Dirección General de Estadística 1917, vii–viii).

The demographic data contained in the yearbook contributed to the measurement of the reality of early twentieth-century Colombia. Colombia recorded a gross birth rate of 31.9 per 1000 inhabitants and a gross mortality rate of 20.9 per 1000 inhabitants, which meant that the population was growing: a constant concern among nations at the time. This was not a minor result for a country torn apart by outbreaks of political violence. The cause-of-death statistics for 1915 were incomplete as 46.2% of the figures were in the categories "poorly defined deaths" or "unspecified causes of death." However, the country showed strikingly high mortality rates for typhoid fever, smallpox, measles, and whooping cough. As highlighted every so often by physicians and social reformers, these figures gave shape and physiognomy to the precarious health and living conditions of the Colombian population.

Health Statistics

In an attempt to fulfil both international agreements and the pleas of local physicians, the government passed another law in 1916, assigning new statistical duties to municipal bureaucracies. From then on, these units would keep a log of nosographic statistics, including the diagnosis of the illness causing the death, which had to be certified by a physician and recorded according to the Bertillon nomenclature (República de Colombia and Gobierno Nacional 1916). However, as soon as the law passed, exceptions appeared.

In January 1917, for instance, the Norte de Santander Health Board determined that mayors would henceforth be responsible for completing records of monthly births, marriages, and deaths in the municipalities and districts of the department. Birth records would state the sex and legitimacy or illegitimacy of the new-born. Mortality statistics would indicate which illness or accident caused the death, and burial was prohibited without a certificate from the attending physician or from two suitable witnesses. In these cases, the priority was to establish that the death had not been caused by a crime. The shortage of professional physicians in Colombia somehow forced these exceptions, which inevitably affected the precision of data gathering procedures. In these adjustments, civil authorities rarely depended only on themselves. To keep their logs, mayors usually had to ask for the help of parish priests, from whom they retrieved statistical information.

In June 1918, the Central Council of Hygiene promulgated an administrative act to regulate the country's nosographic statistics. The act contained five articles. One ordered that every month town mayors and police chiefs send death statistical records to their respective health board. If there were physicians in the area, deaths had to be classified by illness or cause. Likewise, directors of healthcare facilities and public offices, as well as garrison and prison physicians, would have to submit monthly statistical reports of their establishments, listing illnesses according to the Bertillon nomenclature. Municipal health offices and municipal health commissions had to request data on patients with infectious diseases from practising physicians in their municipalities. There were also protocols to order the shipment of regular data to the Council of Hygiene's headquarters. A final article established a fine of five Colombian pesos for any violation of these norms (Junta Central de Higiene 1918). In several of these measures, Colombian authorities followed the recommendations of the international health and sanitation conferences the country had joined to place Colombia within the international arena. But at the same time, they perfectly dovetailed with the wider script of the government and its attempts at transforming the production of statistics.

Four months later, the government passed a law replacing the Central Council of Hygiene with an office called the National Board of Hygiene (*Dirección Nacional de Higiene*). This board was responsible for coordinating, supervising, and regulating public and private health across the nation. Surprisingly, the norm sought to lessen the burden on the public treasury, since it turned departmental administrations into the economic supporters of public health services. But not surprisingly, the savings obtained with the law were scarce compared to the obstacles it brought about. Most departments had to reduce salaries and eliminate jobs. In turn, municipalities refused to pay for the services of municipal health offices, arguing that these tasks were also a responsibility of departmental administrations. As a result, the activities of the municipal offices were restricted or suspended, and the organisation of the national public health administration was weakened (García 1919, 299–300).

Two years later, the National Board of Hygiene passed a new bylaw on health statistics. It obliged all medical practitioners to provide administrative or health authorities with information on patients with infectious diseases excluding individual identities, so as to uphold confidentiality. This law also included information on doctor-assisted deaths where it was mandatory to state the cause following the Bertillon nomenclature. The same criteria applied to child mortality and morbidity data. Likewise, it ordered mayors to gather the data provided by doctors and parish priests on births and deaths and submit it on a quarterly basis to the National Board of Hygiene (Dirección Nacional de Higiene 1932). As it was usual at the time, the means given for enforcing these protocols were inadequate, and this limitation prevented the norms from being totally effective. However, they played a critical role at international level since they guaranteed Colombia's commitment with fighting infectious diseases and moving towards global standards. Indeed, at the Sixth Pan-American Sanitary Conference held in 1920, Colombian delegate Pablo García Medina read a deliberately optimistic report. According to his account, Colombia had fulfilled the requirements put forward by international health conferences and conventions. This portrait, however, concealed an urgent need to promote Colombian agricultural products, especially coffee, within the international market, and to show that infectious diseases did not exist or were under control (Hernández et al. 2002).

During the 1910s, the insertion of Colombia into the international trade system continued to be a national priority. The production of statistics shows that the country followed the route of modern nations, regardless of how partial and incomplete these efforts were. Although it seemed a marginal outcome, the publication of the *Anuario Estadístico 1915* made it possible, for the first time, to make comparisons with other countries. In political terms, the numbers the state could present were far from ideal, but they suggested that despite public health problems, its population was growing.

Centralisation Above, Fragmentation Below

Foreign Experts and the General Accounting Office

In 1923, a delegation from the United States travelled to Colombia to reorganise the country's finances and propose a plan for the budgetary execution to sort out the compensation for the loss of Panama in 1903.

This was known as the Kemmerer Mission. Among several proposals, the mission recommended a fiscal reform, as well as the creation of a central bank and a general accounting office. The creation of this new administrative department followed the Anglo-Saxon model of the Comptroller General, applied for years in England and implemented in the United States in 1920. Alberto Schlesinger, then director of the General Board of Statistics, reorganised the national statistical appartus alongside the mission (Estrada 2015, 174–175).

In the exchanges between the foreign mission and its Colombian counterpart, Schlesinger went over the difficulties his office faced in producing and publishing its reports. Among them, Schlesinger stated, departmental directors used to appoint staff according to political criteria. Due to the conservative/liberal political division, the appointment of statistical officers was often the result of political affinities rather than according to their qualifications. During the first decade of the twentieth century, especially after the presidency of Rafael Reyes (1904–1909), conservatives had opened some room to the liberal opposition through reforms to the electoral system. Although this did not compromise their hegemony, as they continued controlling elections and allocating public post as they pleased, it did allow liberals to accumulate some bargaining power. All this contributed to a certain diversification of the political landscape, but had scarce impact on the daily functioning of the state and on the professionalisation of its bureaucracies (Melo 2020, 179–183). Moreover, wages tended to be low, further hindering the recruitment of qualified employees. Under such circumstances, the consolidation of a national statistical bureaucracy seemed remote (Kemmerer 1994, 199–201).

Following the mission's recommendations, the Colombian government created the General Accounting Office (*Contraloría General*) in 1923. Consequently, several sections of the Ministry of Finance, including the General Board of Statistics, were incorporated under the purview of this new office. Its primary role was to supervise the management of public funds, compile national accounts, and centralise official statistics (República de Colombia and Ministerio del Tesoro 1923). The centralisation of statistics appeared to be an essential reform for the country. In the words of the Kemmerer Mission:

> Statistics reveal essential information on the nation's resources, trade, industries, transport, finances, and health problems. Thanks to statistics, the Colombian people can learn about their economic situation and follow

> the trade and industrial progress of the nation. For foreigners seeking to do business or invest in Colombian assets, official statistics are the primary means they have to form an opinion of the country. (República de Colombia 1923, 145, author's translation)

Thanks to this reform, departmental offices reliant on local authorities and still outside the control of the General Board of Statistics were removed. In parallel, civil servants used the General Board of Statistics detachment from the Ministry of Finance and its incorporation into the General Accounting Office to transform the image of the statistical bureau, deeply entrenched in the public view with merely fiscal and military aims. From then on, the General Board of Statistics tried to carefully follow international guidelines for the future production of demographic and health statistics.

In 1924, the Seventh Pan-American Sanitary Conference was held, and the first Pan-American Sanitary Code was approved. The primary objective of this code addressed the joint adoption of measures to improve public health through the promotion of international cooperation to prevent the introduction and propagation of infectious and transmissible illnesses and to standardise the collection of statistical information on morbidity and mortality (Oficina Sanitaria Panamericana 1924). As part of this changing scenario, in 1925 a new law was passed devoted to reorganising public health administration and to readdressing the issue of demographic and nosological statistics in Colombia. The law eliminated the National Board of Hygiene, which was replaced by the National Board of Hygiene and Public Assistance (*Dirección de Higiene y Asistencia Pública*) focused on hygiene and charity. The new board established a section in charge of collecting vital statistics. Operationally, the Vital Statistics Section still lacked the administrative and technical resources to fully assume its role, so its inauguration was suspended until 1927. Physician Roberto Concha was appointed as head of the section. He had studied statistics for a year in the United States financed by the Rockefeller Foundation (Estrada 2015, 314–341). Nevertheless, the General Board of Statistics did not lose control of crucial statistical data, like births, marriages, and mortality, but data on diseases and causes of death were entrusted to the Vital Statistics Section.

In 1928, a new decree was issued to overcome the reliability issues of the national nosographic statistics, resulting from the lack of university-trained physicians, which tended to distort the information recorded in death certificates. The decree included a new measure: if there were no licensed physicians around, the certificates would indicate that the deceased had died without medical assistance and, therefore, there was no diagnosis concerning the cause-of-death. This decree also ordered local health authorities to report on infectious diseases, flagging up the importance of such statistics for the national legislative agenda. Surveillance for infectious diseases will be, then, a core public health task, essential to prevention and control efforts. The decree also sought to establish a records-inspection process under the supervision of top-ranked health authorities. Specifically, it arranged that the municipal and departmental statistical offices would have to follow the instructions of the National Board of Hygiene and Public Assistance, and were prohibited from publishing or sharing information, as established by international conventions, the Health Committee of the League of Nations, and other similar international organisations, without the approval of the Vital Statistics Section (República de Colombia 1928). Thus, both the production and the circulation of health statistics were regulated.

The National Statistical Board

Also in 1928, the General Accounting Office was restructured. It was reorganised into three sections: the Administrative Office, the Accounting Section, and the Statistics Section. Within the Statistics Section, the reform established the creation of a new sub-section in charge of conducting the census, until then carried out by an office not attached to the General Board of Statistics (Contraloría General de la República 1928). The bipartisan struggle, ongoing for decades, offered a background for frequent disputes concerning which authority—central/national or departmental—had control over each office and its bureaucracy. Creating an office dedicated to conducting the census reveals the will of the government to overcome such struggles and organise the collection of national statistics in an independent bureau.

In 1934, the National Hygiene Department (*Departamento Nacional de Higiene*) made a new attempt at reorganising statistics. The idea was backed by the national government, which decided that this reorganisation would be single-handedly responsible for monitoring the country's demographic and nosographic statistics (República de Colombia 1934). However, the effort to gather demographic and nosographic statistics through this department did not last, because in 1935 the General

Accounting Office introduced a more ambitious reform. As Table 6.2 shows, the General Board of Statistics was transformed into the National Statistical Board (*Dirección Nacional de Estadística*), which would comprise all the existing statistical sections in the ministries, as well as the Vital Statistics Section in the National Board of Hygiene. The National Statistical Board was in turn divided into nine sections: General Directories and Publications; Mechanical Tabulation; Demographic and Health Statistics; Foreign Trade; Tax and Administrative Statistics; Industry, Transport, and Labour Statistics; Agricultural Statistics; Penal, Penitentiary Justice, Electoral, and Religious Statistics; and Cultural Statistics.

This reform also created a Technical Statistical Department (*Departamento Técnico Estadístico*) to oversee the norms according to which all statistical research would be conducted. The department also had to perform technical supervision and oversee the informational sources, as well as carry out the study and analysis of the results. The reform

Table 6.2 Institutional changes in the Colombian statistical system (1906–1952)

Period	*Name of the statistical agency*	*Location*	*Directors*
1906–1923	General Board of Statistics	Ministry of Finance	Vicente Larra R (1906–1909) José Joaquín Caicedo R (1912–1913) Julio H. Palacio (1914) Rafael Jiménez Triana (1915–1917) Alberto Schlesinger (1917–24)
1923–1934	General Board of Statistics	General Accounting Office	Belisario Arenas Paz (1924–1928, 30) Luis Medina R (1928–1929) Valentín Ossa (1929, 1930) Marco Tulio Gómez (1934)
1935–1952	National Statistical Board	General Accounting Office	Marco Tulio Gómez (1935) Juan de D. Higuita (1936–1941)

Source Author's elaboration

aimed at reducing competition- and authority-related conflicts in certain departmental statistical sections where local authorities had appointed the director, who would become an employee of the General Accounting Office (Lleras Restrepo 1938, 10–23). These measures tried to neutralise the common rivalries between central and local powers, which affected many sectors of Colombia's public administration and diminished their ability to perform their tasks.

But the reforms did not stop there. In September 1936, the Centre for Statistical Study and Coordination (*Centro de Estudios y Coordinación Estadística*) was established to support the centralising efforts conducted by the National Statistical Board. This unit was responsible for designing and coordinating a modern statistical research programme. The standardisation of statistics was necessary in order to produce figures that could be compared across the country's departments, but also to meet international standards. The Centre for Statistical Study and Coordination was also in charge of supervising the selection and training of civil servants. This was a decisive move since most of the sector's employees were untrained. Indeed, the previous year the General Accounting Office had offered special courses to enhance the skills of statistical employees. Following the first course on demography, held between October and November of 1935, some members of the course established the Colombian Demographics Society (*Sociedad Colombiana de Demografía*) and began to publish the journal *Demografía colombiana* (Colombian Demography). Albeit the organisation was short-lived, its brief existence proved the aspiration to tackle Colombian statistical problems by technical means (Guthardt 1936, 1–9).

The General Accounting Office also adopted a new regulation to produce demographic and nosologic statistics in Colombia. The bylaw indicated which types of registers departmental statistical boards should fill in to register births, marriages, deaths, and stillbirths. Cause-of-death forms were only to be completed in municipalities by qualified physicians, and the causes would only be recorded if the death certificate was accompanied by a medical certificate (Contraloría General de la República and Dirección Nacional de Estadística 1937).

In practice, the collection of mortality-related demographic information had two obstacles. The first was the regular provision of death certificate registration books by municipal authorities. The second was the authentication of the medical certificate as part of the death certificate. Over the years, the number of monthly reports sent by municipalities gradually rose. While the General Accounting Office received only 62.3% of the reports in 1934, that rate reached 100% in 1937. This increase shows that somehow the authorities managed to establish record-keeping routines, although it did not mean that all deaths in the country were reported properly. Indeed, the under-developed organisation of public administration at the municipal level in several departments led statistical authorities to believe that over 20% of deaths went unreported (Lleras Restrepo 1938, 68–69).

The authentication of medical certificates was essentially impossible to overcome for statistical staff. Since 1916, and according to the law, the verification of a death by a physician was mandatory across the country, except for places without certified professionals. In 1937, however, the exception was still the rule. Out of the 806 Colombian municipalities, 35.5% had a physician and 30% had a hospital. The total number of medical professionals in the country was 1,512 (1 per 5,731 inhabitants), out of which 825—54.6%—practised in departmental capitals. The country had only three medical schools in Bogotá, Medellín, and Cartagena (Velasco Cabrera 1938). The scarcity of medical personnel in Colombia implied that the General Accounting Office could not publish nosographic statistics for the entire country for several years (Lleras Restrepo 1938, 70–71).

In pursuing the centralisation of statistics—and five decades after the Catholic Church took full control of civil records—in 1938 the state established the Civil Status Registry. From then on, notaries became responsible for keeping such register, and in municipalities without notaries, the task fell in the hands of the mayors. Therefore, the registry certificates they issued became the state's primary form of documentation, while the records of ecclesiastical origin would begin to be viewed as supplementary (República de Colombia 1938). The account of the implementation of this new system and its implications for the production and circulation of demographic statistics in Colombia is yet to be written.

Conclusion

Looking at the interplay between political regimen and statistical system, this chapter studies the obstacles Colombian conservative authorities faced in establishing a centralised network of knowledge production between 1886 and 1930. These obstacles did not affect the collection of data on trade, which seemed to flourish without direct state guidance, but were decisive in defining the trajectory of population and health statistics. Both fields ended up drawing a scene of increasing institutional divergence under a political system that tried unsuccessfully to move towards the opposite direction. That was the paradox of the Colombian conservative political project.

The relative position of the state *vis-à-vis* other actors, its administrative coherence, and the strength of its bureaucratic penetration over the territory were crucial factors in the development of a national statistical system. The history of Colombian statistics in the transition from the nineteenth to the twentieth century is a useful case to test how the weaknesses in these factors had structural effects on the production of official knowledge, regardless of the institutional design or the physiognomy the state adopted. The efforts to instal the Bertillon system or collect reliable data on causes of death are indicative of the mismatches that take place when central authorities lack enough power to organise a vast and efficient network of data gathering. If one adds to this the instability derived from the successive reforms of the agencies in charge of the production of statistics, the absence of properly trained personnel—not only of statisticians, also of physicians, who were key for the production of health statistics—financial shortages, and the consequences of political violence—such as the aftermath of the Thousand Days' War—it is not difficult to grasp why the conservative statistical project did not materialise properly.

Likewise, the signing of the Concordat of 1887 with the papacy is also a factor that should be considered. Even though Colombian history is not an exception regarding the role the Catholic Church bureaucracy played in the production of statistics in the absence of a secular and autonomy bureaucracy, it is particularly revealing of how for the state the pragmatic imbrication of both powers proved useful in the short term but became somehow contradictory in the long term when considering the centralising goals of the conservative policies. The late creation of a civil registry system needs to be considered as it shows to what extent the choices that allowed to surmount contingent state limitations ended up naturalising prosthesis that delayed the consolidation of an autonomous infrastructural power.

In Colombia, the production of demographic statistics attuned to international standards was a phenomenon of the second half of the twentieth century. However, it is impossible to detach that era from the previous efforts conservative authorities made to get closer to global statistical organisations, whose standards had a decisive impact on the later configuration and restructuring of national health agencies and the standardisation of Colombian statistics. Albeit the attempts at centralising the production of this knowledge failed, they served as examples of the pitfalls and deficiencies agencies like the General Accounting Office had to overcome.

References

Arboleda, Henrique. 1899. "Circular sobre la adopción del sistema de Bertillon." *Diario Oficial*, July 18.

Arizmendi, Ignacio. 1989. *Presidentes de Colombia: 1810–1990*. Bogotá: Planeta Colombiana Editorial.

Atehortúa, Adolfo L., and Humberto Vélez. 1994. *Estado y fuerzas armadas en Colombia: 1886–1953*. Cali: Pontificia Universidad Javeriana.

Beaud, J.-P., and J.-G. Prévost. 2010. "L'Histoire de la statistique canadienne dans une perspective internationale et panaméricaine." In *Estadisticas Nas Americas. Por uma agenda de estudos históricos comparados*, edited by Nelson de Castro Senra and Alexandre de Paiva Rio Camargo. Rio de Janeiro: IBGE. https://biblioteca.ibge.gov.br/visualizacao/livros/liv44323.pdf. Accessed 28 September 2020.

Castañeda, Gabriel J. 1899. "Estadística Nacional [Informe sobre la conveniencia de adoptar el sistema de Bertillon]." *Diario Oficial*, July 28.

Charry, Arturo. 1954. *Desarrollo histórico de la estadística nacional en Colombia*. Bogotá: Departamento Administrativo de Planeación.

Contraloría General de la República. 1928. "Nueva reorganización de la Contraloría." *Boletín de la contraloría general de la República* 2 (9): 4–6.

Contraloría General de República y Dirección Nacional de Estadística. 1937. *Cartilla de estadística*, Vol. I Bogotá: Imprenta Nacional.

———. 1938. *Estadísticas demográfica y nosológica 1936*. Bogotá: Imprenta Nacional.

———. 1946. *Anuario General de estadística 1945*. Bogotá: Imprenta Nacional.

Desrosières, Alain. 1993. *La politique des grands nombres: histoire de la raison statistique*. Paris: Editions La Découverte.

Dirección Nacional de Higiene. 1932. "Resolución número 90 de 1920 (septiembre 9) sobre estadística sanitaria en la República." In *Compilación de las leyes decretos, acuerdos, y resoluciones vigentes sobre higiene y sanidad en Colombia*, Vol. I Bogotá: Imprenta Nacional.

Estrada, Victoria. 2015. *La valeur des chiffres: La production et les usages des statistiques démographiques et de santé publique en Colombie 1886–1947*. PhD dissertation, École des Hautes Études en Sciences Sociales.

———. 2017. "¿Cuántos somos? Una historia de los censos civiles y de la organización de la estadística en Colombia en la primera mitad del siglo XX." *Historia Crítica* 64: 141–160. https://doi.org/10.7440/histcrit64.2017.08.

Estrada, Victoria, and Jorge Márquez. 2019. "Defensa de los derechos adquiridos: luchas y albures del ejercicio de la homeopatía en Colombia (1905–1950)." *História, Ciências, Saúde-Manguinhos* 26 (4): 1355–1372. http://dx.doi.org/10.1590/s0104-59702019000400019.

Estrada Victoria, Oscar Gallo, and Jorge Márquez. 2016. "Retórica de la cuantificación: tuberculosis, estadística y mundo laboral en Colombia, 1916–1946." *História, Ciências, Saúde-Manguinhos* 23: 277–299. http://dx.doi.org/10.1590/S0104-59702015005000013.

García M., Pablo. 1919. "Informe del Director Nacional de Higiene al Ministerio de Instrucción Pública." *Revista de Higiene* 9 (115): 279–310.

García, Juan C. 1980. "La medicina estatal en América Latina, 1880–1930." *Revista Latinoamericana de Salud* 1: 70–110.

Guthardt, Emilio. 1936. "La sociedad colombiana de demografía." *La demografía colombiana* 1 (1): 1–9.

Hermberg, Paul. 1941. "Las actividades estadísticas de Colombia." *Anales de Economía y Estadísticas* IV (8): 21–40.

Hernández, Mario et al. 2002. *La Organización Panamericana de la Salud y el Estado colombiano. Cien años de historia 1902–2002*. https://www.paho.org/col/index.php?option=com_docmanandview=downloadandcategory_slug=publicaciones-ops-oms-colombiaandalias=61-la-ops-y-el-estado-colombiano-cien-anos-de-historiaandItemid=688. Accessed 10 June 2020.

Junta Central de Higiene. 1887. "Junta Central de Higiene. Reglamento." *Revista Médica de Bogotá* 116: 769–773.

Junta Central de Higiene. 1918. "Acuerdo número 39 de 1918 sobre la estadística nosográfica de la República." *Gaceta Departamental de Antioquia*, September 19.

Kemmerer, Edwin. 1994. *Kemmerer y el Banco de la República: diarios y documentos*. Edited by A. Meisel Roca, A. López Mejía, and F. Ruiz. Bogotá: Banco de la República,

Lleras Restrepo, Carlos. 1938. *La estadística nacional: su organización, sus problemas*. Bogotá: Imprenta Nacional.

Márquez, Jorge. 2005. *Ciudad, miasmas y microbios: la irrupción de la ciencia pasteriana en Antioquia*. Medellín: Editorial Universidad de Antioquia.

Martínez, Abel F., and Samuel A. Guatibonza. 2005. "Cómo Colombia logró ser la primera potencia leprosa del mundo. 1869–1916." *Colombia médica* 36: 244–253.

Melo, Jorge Orlando. 2020. *Colombia: una historia mínima*. Bogotá: Crítica.
Obregón, Diana. 2002. *Batallas contra la lepra: estado, medicina y ciencia en Colombia*. Medellín: Universidad Eafit.
Oficina Sanitaria Panamericana. Código sanitario panamericano 1924. Available at: http://www.paho.org/spanish/sanitarycode.htm. Accessed 8 Mar 2020
Otero D'Costa, Enrique and Gerardo Arrubla. 1935. "Reseña sobre la historia de la estadística en Colombia." *Boletín de historia y antigüedades* 22 (249–250): 254–260.
Prieto, Fabián. 2005. "Una anatomía de la población colombiana: la técnica estadística en Colombia y el levantamiento del censo de población de 1912." *Memoria y sociedad* 9: 55–67.
República de Colombia. 1889. *Decreto número 476 de 1889 (27 de mayo), orgánico del Ministerio de Fomento*. Bogotá: Imprenta de Vapor de Zalamea Hermanos.
———. 1923. *Leyes presentadas al gobierno de Colombia por la Misión de expertos financieros americanos y exposición de motivos de éstas*. Bogotá: Editorial de Cromos.
———. 1928. "Decreto número 580 de 1928 (29 de marzo) por el cual se reglamenta la declaración de las enfermedades contagiosas." *Diario Oficial*, April 17.
———. 1934. "Decreto número 505 de 1934 (marzo 8) por el cual se aprueba la resolución número 34 del Departamento nacional de higiene que reglamenta la estadística demográfica y nosológica de la nación." *Diario Oficial*, March 27.
———. 1938. "Por la cual se dictan algunas disposiciones sobre registro civil y cementerios." *Diario Oficial*, June 15.
———. 1946. *Anuario General de Estadística 1945*. Bogotá: Imprenta Nacional. VII.
República de Colombia and Consejo Nacional Legislativo. 1888. "Ley 29 de 1888 (22 de febrero) que concede una autorización al Gobierno." *Diario Oficial*, February 27.
República de Colombia and Dirección General de Estadística. 1907. *Estadística Anual de la República de Colombia*. Bogotá: Imprenta Eléctrica.
———. 1917. *Anuario Estadístico 1915*. Bogotá: Imprenta Nacional.
República de Colombia and Gobierno Nacional. 1887. "Ley 57 de 1887 (abril 15) Sobre adopción de códigos y unificación de la legislación nacional." *Diario Oficial*, April 20.
———. 1892. "Ley 34 de 1892 (21 de octubre) aprobatoria de una Convención adicional al Concordato del 31 de diciembre de 1887." *Diario Oficial*, November 9.

———. 1916. "Ley 66 de 1916 (diciembre 14) Por la cual se organiza la lucha contra la tuberculosis, y se adiciona y reforma la marcada con el número 84 de 1914, sobre higiene pública y privada." *Diario Oficial*, December 21.

República de Colombia and Ministerio de Fomento. 1893. "Reglamento número 1 para el establecimiento de la Estadística nacional y para el servicio de las Oficinas de este Ramo, creadas por la Ley 107 de 1892." *Boletín trimestral de la estadística nacional de Colombia* 3: 4–5.

República de Colombia and Ministerio de Gobierno. 1906. *Reglamento interno del Ministerio de Gobierno.* Bogotá: Imprenta Nacional.

República de Colombia and Ministerio del Tesoro. 1923. "Ley 42 de 1923 (julio 19) sobre la reorganización de la contabilidad oficial y la creación del Departamento de la Contraloría." *Diario Oficial*, July 26.

República de Colombia, Contraloría General de República and Dirección General de Estadística. 1938. *Anuario General de Estadística 1937*. Bogotá: Imprenta Nacional.

Rey, María F. 2008. "La educación militar en Colombia entre 1886 y 1907." *Historia Crítica* 35: 150–175.

Rodríguez, Jorge. 1928. *Lecciones de estadística*. Medellín: Imprenta Oficial.

Troconis G., Fernando. 1912. *Apuntaciones sobre la mortalidad infantil en Bogotá. Tesis Facultad de Ciencias naturales y medicina*. Bogotá: Imprenta J. Casis.

Velasco Cabrera, Benigno. 1938. "La sanidad en Colombia." *Boletín de la Oficina Sanitaria Panamericana* 17 (3): 197, 201.

Vidales, Luis. 1978. *Historia de la estadística en Colombia*. Bogotá: Banco de la República.

Warming, Sophus H. 1892. "Informe del inspector." *Boletín trimestral de la estadística nacional de Colombia* 1: I–XX.

———. 1893. "Registro civil." *Boletín trimestral de la estadística nacional de Colombia* 2, 1893, 3–60.

CHAPTER 7

Socio-political History of Latin American Statistics: A Bibliographical Essay

Claudia Jorgelina Daniel, Cecilia T. Lanata-Briones, and Andrés Estefane

This chapter scrutinises the existing academic research that focuses on the socio-political history of Latin American statistics. Its aim is to produce an account, a report, or inventory that assesses the contribution that since the end of the twentieth century the study of Latin America has made to the global knowledge on the history of quantification as well as to the role of statistical reasoning in the development of representations about

C. J. Daniel
Centro de Investigaciones Sociales, Consejo Nacional de Investigaciones Científicas y Técnicas, Buenos Aires, Argentina
e-mail: cdaniel@ides.org.ar

C. T. Lanata-Briones (✉)
Department of Economics, University of Warwick, Coventry, UK
e-mail: cecilia.lanata-briones@warwick.ac.uk

A. Estefane
Independent Researcher, Santiago, Chile

C. T. Lanata-Briones et al. (eds.), *Socio-political Histories of Latin American Statistics*, Studies of the Americas,
https://doi.org/10.1007/978-3-030-87714-9_7

the social and economic world. This assessment makes visible the thematic and methodological diversity of the studies on Latin American countries, with special emphasis on the research on the social and political history of statistics. The knowledge accumulated over the last decades is evidence of the robust trajectory of regional statistical production. The contributions of the socio-political history of Latin American statistics overflow and challenge the old, although still valid, assumptions of scientific diffusionism. The unique historical configurations of this diverse region explain, from the outset, the differences detected in the reception and adaptation of statistical knowledge. Despite national specificities, statistical research has been a constitutive part of the political and social processes of the region. It is an indisputable fact that statistics have operated as knowledge implied in, and not only as a mere instrument to measure, the most significant transformations of Latin America, in particular regarding ethnic, economic, and demographic dynamics. Notwithstanding the fact that today there is an active network of researchers who are producing 'the history of Latin American statistics', the region's contribution to the history of the discipline is still relatively unknown at an international level.

As indicated in the introduction of this edited volume, this lack of difussion explains our interest in bringing these works to audiences unfamiliar with the studies on Latin American statistical knowledge production. This chapter—which can serve as a guide for those interested in the socio-political history of statistics as well as for those interested in Latin America—orders and synthesises the central themes of the bibliographical corpus that is presented at the end of the chapter. Although this list is not exhaustive and this compilation is far from being the definitive view of that trajectory, both inputs—this chapter and the reference list—articulate a representative approximation of the main trends that studies of or about the region have followed. The bibliographical corpus only lists published works, as well as PhD and MSc dissertations.

This chapter is divided into three sections. The first frames the initial interest in the socio-political history of Latin American statistics, highlighting some pioneering texts. This section proposes an interpretation of the political and social processes that determined the paths followed by these first enquiries. The second section, divided into four organising nuclei, examines the phase of 'proliferation' of these studies, noting the various approaches, their specificities considering regional particularities, and the analytical potential they offer as a corpus. The third and final section identifies the issues that are emerging and that could constitute a new stage in this trajectory.

Numbers and Institutions. The First Socio-Political Histories of Latin American Statistics

The socio-political historical approach to statistics is a relatively novel perspective in Latin America that has been configured within the field of social sciences under the wing of social constructivism. However, a series of studies that had already explored production patterns and debates around the political utility of quantitative knowledge nourished the origins of this approach. Between the 1960s and 1970s, a systematic concern for Latin American statistics generated a surge in studies that scrutinised the robustness of the available figures, particularly those of official nature.

The accelerated changes in the productive structures of the Latin American economies and the subsequent significant social and demographic transformations generated the urgent need to look back at the available statistical heritage and assess the possibilities of building up-to-date knowledge on these foundations. Responding to the mirages of modernisation and reclaiming the role that social sciences could play in this journey, political currents related to state planning demanded knowledge that meant to test the consistency of the region's statistical history (Vessuri 2006). Such scrutiny took place within the framework of a long cycle of political tensions expressed by the continental settings of the global Cold War (Pettinà 2018). Amid a revolutionary and counterrevolutionary onslaught, there was a proliferation of questions concerning the economic, social, and political conditions that were at the base of the transformation projects in dispute. Statistics became a core instrument for understanding these conditions. The theoretical and intellectual drive that stirred the Latin American scientific landscape during those decades also had to be considered. This momentum materialised in the production of a 'situated knowledge' [*conocimiento situado*] and committed to the needs of a region that was reluctant to define itself as the simple periphery of a bipolar world (Devés 2003; Lozoya 2020). Statistics were thus integrated into the toolbox of essential knowledge necessary to intervene properly in this situation.

The re-examination of Latin America's statistical heritage took various paths, such as the generation of catalogues and the identification of gaps and areas for strengthening official statistics, as well as proposals for coupling and adjusting demographic, social, and economic measurements following international guidelines (Ward 2004). Of course, these assessments of official statistics had normative biases. Moreover, those biases were projected onto the historical reconstructions of the statistical practice, which symptomatically did not focus only on the republican era, but also on colonial cadastres and on counting practices that focused on people and resources that were described as 'proto-statistical' operations (Gómez 1970; Urrutia and Arrubla 1970; Peachy 1976; Marcilio 1977; Mamalakis 1978; Villamarín and Villamarín 1981).

In a revealing parallel to the expectations projected on this science and like the turn experienced after the Second World War by the statistical agencies of the central capitalist countries, Latin American national statistical offices progressively ceased to be bureaus to constitute themselves as institutes. The aim was to take a symbolic distance from the old idea of bureaucracy, so as to link its tasks to the academic work carried out inside and outside national universities, and gain prestige. As part of the same reform, in the 1960s and 1970s statistical offices in countries like Argentina and Brazil began to be housed in ministries or planning agencies. The offices redefined their role within the state as information production hubs for national economic development plans. From then on, they would adopt an increasingly academic-scientific mode of production. It is not by chance that these new institutional links fostered a new type of disposition by academics and specialised researchers, both towards them and their history.

Indeed, the progressive expansion, diversification, and professionalisation of higher education institutions in several Latin American countries (Brunner 1990; Balán 2013) was the context in which new uses for statistical outputs began to take shape. The modernisation and internationalisation of social sciences also contributed to this process. Beginning with the 1950 and 1960 waves, national population censuses ceased to be understood as mere state measurement instruments to become valuable tools for social research. It is still unclear the extent to which the data produced by national statistical institutes was used systematically by Latin American governments for long-term planning. However, the institutional modifications under way were decisive for the visibility and public legitimacy of statistics. Referring to the Brazilian Institute of Geography and Statistics (*Instituto Brasileiro de Geografia e Estatística*, IBGE), Simon Schwartzman claims:

> But the planning imaginary had deep consequences for the internal organisation of the office [the IBGE]. Now each research line could be said to have a definite place in a coherent picture, and could not be easily challenged or changed. As long as the planning imaginary retained its appeal, the office's legitimacy would remain intact. (1997, 7)

In this scenario, from various subfields of history—such as labour, gender, economic, and demographic history—the need arose to examine quantitative sources from a critical perspective. The shift was based on a certain distrust of the interpretations that had been made of their results, while at the same time it was a warning regarding the use of that information. Criticisms also reached demography, a discipline that showed different degrees of maturity depending on the country, but that had advanced towards a certain convergence after the creation in 1957 of the Latin American and Caribbean Demographic Centre (*Centro Latinoamericano y Caribeño de Demografía* or CELADE) (CEPAL 2009). Thus began a careful review of the methodologies applied when conducting population censuses, with a growing participation of researchers from humanities and social sciences.

The question of trust in numbers imposed itself as an unavoidable horizon when estimating the degrees of precision of the measurements. But there was also an interest in approaching the representations of certain social phenomena—although still under the binarism accuracy/distortion—and in looking into the effects of the 'invisibility' of certain populations (Sánchez-Albornoz 1977; Recchini de Lattes and Wainerman 1979; Recchini de Lattes 1982; Arrext et al. 1983). Given the purposes of this chapter, these works reveal a significant turning point: the replacement of the vision of statistics as a *means* for social analysis by an understanding of them as *objects* of study. These investigations still displayed resentments regarding the excessive relativisation of numbers, but without a doubt they began to shift the trend.

The leap forward from those initial critical approaches to the understanding of statistical data as socially constructed objects occurred in the 1990s. Without necessarily responding to a pre-established research programme, various works began to show an interest in studying what historian Nelson Senra (2005) successfully named the "intimacy" of statistics. Population censuses were the first point of attraction for this new approach. In some cases, this shift pointed towards the evaluation of the institutional capacities of states to count their population, while others explored the relational dimension of statistics, analysing their role in the forging of national imaginaries. This occurred for Peru (Gootenberg 1991), Argentina (Otero 1997–1998, 1999; González Bollo 1999a, b), Brazil (Senra 1997; Rodrigues Botelho 1999), Mexico (Lomnitz 1999), and Chile (Hutchison 2000; Pinto 2002–2003), to name a few examples.

The horizon that these works established aligns with Elizabeth Hutchison's (2000) proposal to examine "the history behind numbers". Although the willingness to submit censuses to a rigorous methodological critique was not completely abandoned—as the handbook of the 'good' social scientist would recommend—in Hutchison's phrase this new qualitative reading of censuses took form, conceiving them as a secondary source. Following Hernán Otero, censuses should be understood "as an intellectual production that testifies to the intellectual and political conceptions of its authors or of an era" (Otero 2006, 34, authors' translation). Thus, these instruments were analysed looking into their socio-historical conditions of possibility and considering the political projects that contained them. The ideological assumptions and the political concerns that census-takers of the nineteenth and early nineteenth centuries applied when listing, classifying, and ordering the population were still unexplored themes, which explains the value of these investigations and the ample range of the research questions.

If these trajectories are traced at the national level, much of these first works focus on the detailed study of specific censuses. These cases became starting points for more substantive generalisations about statistical practices or about the institutional interactions that supported them. An illustrative case is Peru. Within a couple of years, crucial studies informed us about three significant census experiences: 1725–1740, 1860, and 1876 (Chiaramonti 2000a; Pearce 2001; Peloso and Ragas 2001). Including those that had this specific focus, almost all the works on censuses published in the 1990s and some in the early years of the twenty-first century aimed to de-naturalise the use of numbers as fetishes of objectivity and to prevent the mechanical reproduction of the 'lenses' or cultural patterns of contemporary statistical tools (Piza and Rosemberg 1999; Prieto 2005). Benedict Anderson's (1983) classic study of the origin and spread of nationalism, which includes censuses among institutions of nationalist pedagogy, also helped to underpin this shift. Some studies that integrated the censuses in the diversity of statistical outputs generated by Latin American bureaucracies have similar purposes (Schwartzman 1997). The study of the production of sectoral statistics and national yearbooks (Senra 1997) meant an expansion in the range of statistical sources analysed, and this implied the formulation of other questions concerning the structure and objectives of these reports, as well as new standpoints of observation, such as the policy of dissemination and circulation of knowledge. In parallel, the first Mexican studies on the history of statistics during the first half of the nineteenth century focused on the analysis of small scientific communities (Mayer Celis 1999) and were closely inspired by the contemporary discussions in the field of the history of science.

Taken together, these studies forged a crucial framework for understanding the paths that the socio-political history of Latin American statistics followed in the first two decades of the twenty-first century. Without dismissing the particularities due to political and academic national contexts, what happened in this area responded to broad processes that should be considered to understand the insertion of these 'peripheral histories' in the study of the global trajectory of the discipline.

The 1990s entailed profound economic reforms following the guidelines of the Washington Consensus, placing most Latin American countries on the global map of neoliberalism (Escalante 2019; Ruiz Encina 2019). The consolidation of the market as arbitrator and allocator of resources implied a substantive shock for societies in which the state had played a central role as an agent of economic development (Cavarozzi 1996). Due to the profound restructuring of state agencies and to the acute social consequences of these reforms, the state itself became the centre of attention of the local intellectual communities. Statistics emerged as a revealing gateway to understand the trajectory of state action and the meaning its transformations.

To achieve these aims, scholars had to look back at the history of the nation-state, the forging of its legitimacy, and the technical-administrative peculiarities of its bureaucratic and military apparatuses (López-Alves 2000; Dunkerley 2002). But they also had to closely examine the interaction patterns of its agents and representatives with plural societies (Joseph and Nugent 1994). Thus, the study of the obstacles encountered by the Latin American states in their attempt to monopolise the extraction and administration of information on the population and the territory, as well as the analysis of the difficulties to form consistent, predictable, and efficient statistical systems, offered unique clues to think about the Latin American states themselves.

In this context, the penetration of the constructivist approach within Latin American scholars intensified. Here it is convenient to highlight the implications for regional research of the end of authoritarian governments and the gradual recovery of democracy. This meant greater levels of openness, communication, and exchange opportunities for Latin American researchers. The reception of what Corcuff (2013) calls "the new French sociologies" of the 1980s and 1990s and the work of Pierre Bourdieu became decisive for the study of statistics. Bourdieu placed at the centre of his theoretical framework the concern for the social conditions for the production of knowledge about the social world. From there, he encouraged reflection on the practices behind the elaboration of statistics as well as on the categories and classifications that statistics offered to perception.

The echoes in Latin America of the post-Khunian turn, which reformulated the division of tasks between an 'internalist' and an 'externalist' history of science, must also be considered. This turn led to blurring the underlying distinction between technical and social objects. Thus, the interest in Latin American statistics unfolded within the framework of a new epistemological shift that considered the social construction of knowledge. From then on, it would be rare to find research that assumed statistical data as 'given' facts or as transparent information. On the contrary, the new conventional point of view implied that statistical data was the result of a process of knowledge production whose social conditions of possibility could and should be explained. If statistical techniques and instruments were historically and socially situated, then the task of researchers was to reconstruct their contexts.

The socio-political history of Latin American statistics thus incorporated the requirement of reflexivity that characterised the latest trends within social sciences. In turn, it tended to see in the dynamics of the generation of this knowledge one of the ways in which societies represented themselves. Paying attention to how statistics were constructed was also a way of generating a critical view of Latin American societies. This was the opening stage for the remarkable expansion of historical studies on Latin American statistics at the beginning of the twenty-first century.

The Evolution of a 'Symptom'. The Socio-Political History of Latin American Statistics in the Twenty-First Century

In the 2010 edited volume *Estatísticas nas Américas*—a book that continues to be essential to portray the study of statistics in the continent—Otero describes the birth of a specific sub-field among the Argentine social sciences devoted to the study of statistics as a discipline and of statistics as a scientific product. Otero refers to a "historiographic symptom"—a formula that the French historian Jacques Revel used to define his position *vis-à-vis* Italian microhistory—to account for the density of these works. The scene formed by these studies was still very incipient in order to speak of the existence of a "field", but it was active enough to say that they revolved around "common concerns". The medical metaphor was complemented by the fact that specialists from various disciplines came together, although this did not generate an epistemologically *mestizo* space (Otero 2010, 77).

Although Otero referred to Argentina while there were other countries with a stronger record regarding the study of the history of statistics—such as Brazil or Mexico—his metaphor seems appropriate to characterise the Latin American context in the dawn of the twenty-first century, and it could be used as a point of reference to think what has changed since then. Otero (2018) himself, in fact, advances in this line with a recent review also aimed at putting the research on the socio-political history of Latin American statistics on the radar of European and North American specialists. This time covering Latin America as whole, Otero's reconstruction acknowledges there has been substantial progress in the configuration of a regional field, highlighting the existence of collective publications and thematic special issues in journals that bring together representative works about national trajectories (Senra and Camargo 2010; Carvalho Jr. et al. 2011; and the 2017 special issue of the journal *Histoire & Mesure*). He also reminds us of the meetings in international congresses of the scholars that study Latin American statistics (Rio de Janeiro 2009, Salvador de Bahía 2010, Mendoza 2013, São Paulo 2016, and Montreal 2017, to which we can add Santiago in 2019 and the 2021 online conferenced organised by Mexican scholars). Viewed in sequence, these academic gatherings confirm the efforts to deploy a joint agenda. Lastly, Otero discusses the establishment in 2011 of the Association of the Americas for the History of Statistics and the Calculation of Probabilities

(*Asociación de las Américas para la Historia de la Estadística y el Cálculo de las Probabilidades)*, a platform that holds its own journal, *Estadística y Sociedad*, with five issues released between 2011 and 2018.

Otero's account of the changes experienced by this field in the first decades of the twenty-first century suggests the densification of the knowledge about national trajectories. In today's eyes this could be described as a problem and even as a limitation, as Otero does in his most recent review, thereby expressing the influence of the global turn within the historiography. Since that shift, now gravitating, the inertia of continuing to study the past according to the limits of the nation-state has been sharply criticised (Hunt 2015; Conrad 2017; see also van Ittersum and Jacobs 2012; Adelman 2017; Appadurai 2020; Drayton 2020). Given its links with the historiography on state-building, the study of the socio-political history of statistics is at the centre of this debate.

The criticism of the national emphasis that takes place in these studies has several dimensions. Since its global massification, statistics have been put under strain. On the one hand, there was its universalist identity, derived from its scientific profile. On the other hand, its local identification arose from its use as a tool for government and administration in diverse social formations. Was statistics a global knowledge or a national science? This dilemma is similar to what Nico Randeraad (2010) describes as the theoretical and practical conflicts between cosmopolitan agents formally committed to the ideals of uniformity and neutrality of an emerging science and the pressing demands of smaller 'imagined communities' concerned with adjusting the reception of statistics to the needs of their political systems.

The clash between universalist representations and national approaches has been a constant within the socio-political history of statistics, and it is difficult for those who study it to avoid this constitutive ambivalence. The invitation to combine or to 'overcome' the study of national and state trajectories with comparative, transnational, or global exercises constitutes a great opportunity to look at these processes from new perspectives. However, this cannot dismiss the fact that none of these attempts will be substantive if knowledge of local experiences remains superficial or imprecise. Precisely for this reason, it is pertinent to balance post-national and cosmopolitan anxieties with a weighted evaluation of the paths that the socio-political history of Latin American statistics has effectively taken (*looking at what it has been*) and avoid the drive of comparing its trajectory with impossible ideal types (*thinking of what we would like it to be*). In any case, these are highly problematic gaps for the humanities, the social sciences, and, of course, for the socio-political history of statistics (Brown 2015; Adelman 2017; de Lima Grecco and Schuster 2020; Chernilo 2020).

Without ignoring this tension, in this chapter we propose an understanding of the recent scholarly production following four thematic nuclei: the history of statistical apparatuses and their technical-bureaucratic communities; the history of discourses and ideas, with particular emphasis on the role statistics played in forging national imaginaries; the interplay between domestic and international networks regarding the production and circulation of statistical research; and the socio-cultural history of Latin American statistics that traces the intersections between scientific cultures, technical instruments, and statistics. The works of the twenty-first century can hardly be pigeonholed into a single category. Hence, this is a pragmatic division carried out with the sole purpose of ordering the existing scholarly output. Nonetheless, this division is useful and far more suggestive than a reconstruction that replicates national boundaries, which, strictly speaking, would also be imprecise. Although it is true that most studies focus on national spaces, it is in the study of common problems that connection points have been established. Any revision of the bibliographical corpus listed at the end of this chapter will show that there has been more dialogue than what is commonly accepted.

States, Statistical Apparatuses, and Specialised Bureaucracies

The socio-historical studies on the procedures of statistical objectification and their organisation in institutional spaces within Latin America highlight the role those cognitive technologies played in the processes of state and nation formation. This is, of course, a broader historical problem where statistics tend to play an important role (Centeno and Ferraro 2013; del Castillo 2018; Miller 2020). Not surprisingly, these studies have been linked to the exploration of bureaucratic-institutional factors and in particular with the efforts of the states to lay the foundations of modern infrastructures for the collection of statistical data. In general terms, these studies converge in the thesis that statistical aggregation was at the same time the act of representation of reality and the construction of a political authority.

An important part of the research carried out in recent decades centres on the nineteenth century and in the era of the state-building. That concentration explains the emphasis on institutional factors. The focus revolves around the dynamics of the production of official figures, the organisational structure and the rules followed by the statistical offices, the financial and human resources available to them, and the path of the research agendas. Within this broad research arc, there is, in turn, a persistent dialogue with the problems of social and political history.

Echoing the classic discussion regarding the radicality or gradualness of independence and the collapse of the Spanish Empire, several works study the lines of continuity between the colonial and republican periods, although there are also others that record the loss of practical knowledge or know-how between the two periods. This loss happened in Peru, and it was a direct consequence of the absence of a group of civil servants associated with the central government that was able to send and store information in amounts equivalent to the information flow handled during the colonial period (Ragas 2008, 10). This situation affected the Peruvian statistical apparatus at least until the second half of the nineteenth century, when the state finally managed to establish a network of prefects and sub-prefects. In Chile, on the other hand, the colonial experience influenced the production of statistics during the first decades of the republican period because the intellectuals and bureaucrats who had carried out investigations under the orders of the empire remained active and fulfilled similar tasks in the nascent republic. According to Andrés Estefane (2019), this explains the kinship between colonial topographic descriptions and the first republican research programmes, as well as the persistence of a descriptive-qualitative approach in statistics and a certain continuity in data collection protocols.

Laura Cházaro (2001) makes a similar argument for the Mexican case, pointing out the impossibility of ruptures in terms of statistical ideas and practices after independence. But this did not imply reproducing the illusion of total continuity, since the Mexican civil servants and scientists of the nineteenth century would have re-signified the instruments they inherited and the very idea of population in the light of the emergence of new political discourses. In this case, inheritance did not mean inertia. The Brazilian experience, with its own political chronology, would reinforce this picture. With the completion of the first modern census of Brazil in 1872, Alexandre Camargo (2018) identifies continuity with the estimates and nominative lists that preceded it, despite the rhetoric of rupture and statistical innovation sustained by the imperial authorities. The persistence was evident in the intervention of the enumerators, whose practice of direct observation—still not disciplined by standardised protocols—continued to be understood as a truth factor for designing, applying, and processing questionnaires. The same applied for the restrictions regarding the questions that could be posed about the phenotype and condition of the population, as in the nominative lists.

Studies related to the institutional history of statistics tend to emphasise the difficulties, obstacles, and challenges of organising statistical services in most Latin American countries. One aspect studied relates to the difficulties that geography and territorial integration imposed on the population counts, which seemed to be exacerbated by the imbalances that the federal systems implied for the centralisation and uniformity with which statistical operations were understood. There were also bureaucratic disconnections generated by the struggle between civil and ecclesiastical authorities, central actors in the dispute for the monopoly on demographic information, or the hierarchy and subordination problems experienced by the nascent statistical agencies in their relationship with other state agencies. In these scenarios, operational deficiencies derived from budget constraints and from the poor training of those who served as generators of official statistics were common.

The studies of population censuses and of the national distribution of questionnaires for sectoral statistics have been valuable in this line of research. Likewise, the exploration of local spaces and the interaction between bureaucracies and citizenships have placed the production of statistics as an illustrative arena of the processes of construction of state legitimacy. It is common to find here clues about forms of passive resistance or open indifference among the provincial, state, or departmental bureaucracies, combined with the analysis of the secular mistrust of the population in the face of any type of registration that on some occasions led to notorious episodes of violence.

Several of these problems appear, for instance, in the study of the forging of the Peruvian statistical apparatus and in its most significant census operations throughout the nineteenth century (Peloso and Ragas 2001; Ragas 2008; Chiaramonti 2000a). The same happened in Chile, despite the statistical divergence between the two countries after independence. The institutionalisation stage of the Chilean Statistical Office (*Oficina de Estadística*) between 1843 and 1851 was particularly blocked. The office was questioned from the beginning by the national congress due to the number of financial resources that sustaining it would demand (Estefane 2016). Hernán González Bollo (2007a; 2014) analyses the

Argentine statistical bureaucracy during the second half of the nineteenth century, observing the imbalance created by a variety of statistical offices that configured a dispersed scenario, which he characterises as an archipelago. The difficult process of configuring census statistics in the first half of the twentieth century in Colombia is another case that demonstrates the fluctuations in this trajectory and the importance of looking at them in the light of broader dynamics (Estrada Orrego 2015).

The inertia to analyse national trajectories under the assumptions of incremental institutionalism or using the success/failure binomial based on the 'capacity to count' of Latin American states is overcome by regional studies. The adoption of historical-contingent perspectives neutralises the understanding of the trajectories of statistical apparatuses as failed, incomplete, or distorted experiences with respect to a single path or a predetermined outcome (Otero 2006). Because of that, the study of statistics as a form of administrative knowledge and practice is a fertile field for rethinking institutional problems and socio-political conflicts within Latin America.

The history of statistics shows, for example, new dimensions concerning the secular tensions between civil and religious authorities. Once the studies of the processes of secularisation and laicisation overcame the reduction of the problem to a mere ideological conflict between liberal modernity and Catholic traditionalism, there was an opening up to the consideration of events and processes that depict the political density and the implications for institutions of a conflict with bureaucratic and cultural components, and more grey areas than are usually considered. The intensity of the dispute between the Catholic Church and the state for the control of demographic information on births, marriages, and deaths, for example, is one of the key inputs to understand the rhythms and scope of the configuration of the informational monopoly claimed by Latin American states throughout the nineteenth century. These struggles suggest the centrality of the ecclesiastical bureaucracy, its experience, and social legitimacy, when its state equivalent was still clumsy at collecting information and hence to build an autonomous statistical regime.

During the colonial era, as Nicolás Sánchez-Albornoz points out, "parish priests and *doctrineros* [indoctrinators] knew their parishioners better than the *corregidores* [mayors] knew their people" (1977, 22, authors' translation). During the first century of republican rule, the situation was not substantially different. Civil bureaucracies that understood this initial handicap were more effective in the long run than those that attempted to take parish records by storm. The Brazilian experience is illustrative in this regard, as shown by Mara Loveman's studies on the failed population census of 1852 as well as Camargo's on the census of 1872. Concerning the 1852 enquiry, the government's attempt to introduce a civil registration system and take the first population census ended in a popular revolt in north-eastern Brazil, the so-called *Levante dos Marimbondos*. Part of the problem was the state's strategy of intervening in open opposition and not in instrumental harmony with the Catholic Church, who until then maintained control of demographic records (Loveman 2007a). This situation was very different from what occurred with the 1872 census, as its deployment rested on a careful connection between the ecclesiastical structure and the administrative apparatuses that were patrimonially controlled by the local elites (Camargo 2018, 418). Similar performances and civic-ecclesiastical alliances occurred in other countries, such as Peru and Chile, especially when offsetting the absence or precariousness of state registries (Chiaramonti 2000b; Estefane 2017).

Socio-political statistical studies also illuminate the inexhaustible conflicts between central and local powers that are critical in a region that has followed divergent paths in the degrees of centralisation of state power. Despite statistical activity in Brazil being formally managed by a central office—the General Board of Statistics (*Diretoria Geral de Estatística,* DGE-B)—during the First Republic, Senra (2006) highlights how weak the inter-administrative commitment was between the federal state and the municipal spheres. In a federal country like Brazil, the difficulties of cooperation and coordination between jurisdictions would persist even in the Vargas era. By then, the reform of the Brazilian statistical system itself would be aimed at averting this problem (Senra 2008). Existing research shows the emergence of bureaucratic and administrative conflicts arising from the establishment of national bureaucracies in charge of collecting information. Political conflicts given the centrality of censuses and statistics in general for the definition of representation quotas, the administrative division of the territory, and the distribution of governmental powers at the subnational level were also an issue. González Bollo studies some of them (2010b). Analysing the gap between the 1869 and 1914 Argentine population censuses, he traces the resistance to demographic counts within the state, particularly within the national congress, given its confrontation with the interests of the representatives of the less populated provinces of the interior of the country. Conflicts between central and local powers also took place in Colombia in the framework of attempts to centralise public statistical services starting in the 1920s (Estrada Orrego 2017).

There are also approaches that focus on explaining the role of statistics in critical instances, serving as a bridge between institutions or offering a language to name and measure phenomena to which reference was made inaccurately or with obsolete definitions. Regarding the global crisis of 1930 and the need to measure its effects, two works by González Bollo (2004, 2007a) show for the Argentine case the sophisticated interaction between local statistical management composed by bureaucracies, offices, and publications, and other institutions and areas of public administration. Something similar happened in Chile given the need to understand the conflict between capital and labour in the framework of the 'social question' that led to the establishment in 1907 of the Labour Office (*Oficina del Trabajo*), a crucial milestone for statistics in general and labour statistics in particular (Yáñez 2008). Argentine socio-labour statistics followed a similar trend and were closely linked to the trajectory of the National Labour Department (*Departamento Nacional del Trabajo*) (González Bollo 1999a, 2004, 2012b; Daniel 2011b, c). The works by Cecilia Lanata-Briones (2016, 2021), meanwhile, examine how in Argentina the 1930 crisis helped to reposition and redefine the cost-of-living index.

Along with the establishment of national statistical apparatuses, another focus of interest in recent historiography is the trajectories of the main representatives of local statistics. The analytical strategy of identifying personal and intellectual networks served, for example, to rebuild the existence of a community of statisticians in Mexico in the nineteenth century (Mayer Celis 1999). For Argentina there has been ample research concerning biographies, prosopographies, and studies of the trajectories of crucial individuals. The dialogue between these works establishes three stages: the formation of a specialised bureaucracy, the consolidation of a statistical elite and a generational change during the political expansion promoted by the Argentine liberal state, and the Peronist state (González Bollo 2010a, 2012a; Daniel 2012b; Pereyra and González Bollo 2020; Lanata-Briones 2021).

Certain specificities became clear when explaining the relationship between the formation of statistical apparatuses, the establishment of specialised bureaucracies, and the configuration of communities of experts. Some studies follow the trajectory of key figures that are distinguishable by their influence on local statistical practice (González Bollo

2010c; 2012a). Other works choose to trace the role of the intermediate bureaucratic strata, generally without technical training and with little impact on the local history of statistics, but whose action was essential for learning about information gathering practices in territories far from the nation's capital (Estefane 2012). The notion of "author statistics"—suggested by Otero (2006) to label a specific period or stage in the trajectory of the Argentine statistical apparatus—is closely related to the first set of studies and is used as an interpretive key to analyse other national contexts such as Brazil (Senra 2006–2008; Camargo 2009). The second approach does not have an equivalent reference. However, it is useful to show the practical limits of the exercise of state power in the provinces, as well as to question the assumed subordination of the statistical commissioners to the directives emanating from the country's political centre. Following this idea, Estefane (2012) identifies provincial statistics civil servants as crucial actors in the assembly of a territorial and social network for the generation of information, who acted as translators of the instructions received from the national authority and as encoders of knowledge of influential individuals in each locality.

Taken as a whole, these works explicitly acknowledge the crucial role of the state in the generation of quantitative knowledge about society and the claim of that power to establish itself as the sole producer behind this practice. Unlike what happened in other latitudes, where scientific societies, trade unions, and civil society played an important role in the production of standardised aggregate and sectoral statistics, in Latin America the focus had tended to be on state action or at least in direct reference to its deployment. This raises the unavoidable question regarding the possibility that the state veneration that characterises Latin American historiography is making invisible statistical production practices outside the institutional frameworks of the state.

Statistical Discourses and National Imaginaries

Parallel to its place in the processes of construction of Latin American states, the literature acknowledges that statistics had a crucial role in post-colonial Latin America to imagine the nation, construct it symbolically, and justify it politically. Historical studies carried out along these lines analyse national population censuses as instruments that contributed to the construction of a certain image of the nation. These works make use of the concept of statistical discourse as an interpretive key to the political function acquired by modern censuses and its link with the historical description to establish a genealogy of the nation. Pioneers like Leticia Mayer Celis, Nelson Senra, and Hernán Otero introduce the notion of statistical discourse in the analysis of Latin American statistics. Their works were the first to approach statistics as a cultural text, and to analyse its verbal, matrix, and graphic languages as well as its implicit mechanisms. As Camargo (2015, 93) highlights, the study of the configurations of the statistical discourse constitutes a valuable reading that becomes crucial to enhance a historiography that analyses already very closely examined periods of national histories like the phases of nation-building.

Promoted as true inventories of the nation, national population censuses helped to forge that imagined political community that in independent Latin America meant the homogenisation of very heterogeneous geographic and social realities. As noted by Loveman (2013, 330) "political elites not only confronted the monumental tasks of creating durable bureaucratic structures and expanding the state's infrastructural reach. They also confronted the task of transforming the rhetorical fiction of 'the nation' into a broadly accepted social fact". In the search to generate an amalgamating effect in conflictive and culturally diverse societies, statistics were the tool capable of 'reinforcing' these realities and putting them in front of the eyes of locals and foreigners.

Regarding periodisation, the links between statistics, censuses, and nation building are studied for the first half of the nineteenth century in Mexico (Arrioja Díaz Viruell 2016), and in the case of Brazil for both the imperial period (Rodrigues Botelho 2005; Camargo 2018) and for the republican stage of the oligarchic state (Loveman 2009; Camargo 2015). The analysis of censuses as crucial tools in the complex process of social and political construction of the nation extends even to the middle of the twentieth century as analysed by Raúl Necochea López (2010) concerning the 1940 Peruvian census, and by both Michael Ervin (2009) and Héctor Vera (2017) relative to post-revolutionary Mexico.

Censuses helped create and envisage an image of the nation. This link is established by analysing the thematic interests and silences of the statistical tools, the selection and definition of categories, the ways its results were presented, and the interpretative readings of its findings by census directors. In the Argentine case, Otero (1997–1998, 1999, 2004, 2006) examines the discursive matrix of the first three national population censuses taken in 1869, 1895, and 1914, highlighting their symbolic implications for the statistical construction of the nation during the oligarchic state. The evaluation of the dynamics of concept construction in connection with the intellectual, social, and political context of turn-of-the-century Argentina reveals the continuity or paradigmatic unity of the first three national censuses. For Chile, Estefane (2004) identifies the contributions of the eight population censuses taken during the nineteenth century to the nation-building process in that country, which, similar to Argentina, aimed to homogenise a heterogeneous geographical and social reality. However, taking the census as an event, and at the level of the territorial deployment that this type of registry implies, Estefane also highlights the construction of social loyalties—with the patriotic assistance of the enlightened neighbours of each local community—needed to carry out the census and its incidence in affirming both the symbolic and practical boundaries of the nation.

Some studies aim to elucidate the way in which the national or official statistical system thought about the population and the long-term symbolic effects of this numerical operation. Cházaro (2001) focuses her attention on the images created of the Mexican population, understanding that statistical instruments offered, rather than measurements, true national identities, amalgamating moral and political values behind their apparent rigour and numerical precision. She also demonstrates the coexistence and the differences between counting or enumerating, one by one, the population *vis-à-vis* calculating or estimating the population, both of which were pursued with the same enthusiasm by civil servants and members of scientific societies in nineteenth-century Mexico. Mayer Celis (1999) highlights the creation of an image of a Mexican individual as a typical man, similar to Adolphe Quetelet's theory of the average man. Loveman (2009) underlines the change in the traits of the population considered essential for the progress of the Brazilian nation, from educational aspects towards a gradual racialisation of the DGE-B's institutional definition of progress. In Argentina, the liberal political-statistical paradigm, contrary to the colonial organic conception of the population (Otero 2006, 378), aimed to homogenise a heterogeneous population under the interest of the census takers in accounting for the viability of the nation that was being built.

Just as they were the privileged source of analysis for the historiography at its beginnings, national population censuses became the focus of one of the first systematic attempts to generate a global or general vision of Latin American statistics (Loveman 2014). Rather than being guided by a comparative ambition, Loveman's work is motivated by the need to identify trends shared by groups of countries with some common condition. She reviews how the region's nineteenth-century censuses assumed the existence of an identifiable and demarcated national population to count. Although these countries built and put forward an image of a unified nation, each census implied a different way of handling the ethnic, racial, and cultural differences present in those territories either by exclusion—in Argentina, Chile, and Bolivia—or by assimilation—in Brazil, Cuba, Colombia, Guatemala, and Mexico.

Delving into the relationships between statistics and the national imaginary means putting under the magnifying glass the operations carried out through the census categories in geographically close or bordering countries that did not share identical census grids. While Otero (2006) highlights the methodological minimalism that guided the census-takers to erase the traces of racial differences within Argentina, Camargo (2009) analyses the meanings assumed by racial classification in modern Brazilian censuses up to the 1940. Camargo emphasises the tensions between political expectations and the technical possibilities behind racial classifications, considering the scientific culture and the space in which the producers of official data operated, as well as the positions and the change both in the profile and in the performance of statisticians.

In addition to tracking their topic and subject obsessions, the literature is interested in scrutinising the assumptions and interpretations of census-takers and other statistical civil servants, as the works on Brazil (Rodrigues Botelho 2005; Camargo 2015) and Argentina (Otero 1999, 2004, 2006) show. To analyse Argentina Otero introduces the concept of "statistical ideology"—defined as a set of representations about the social reality implicit in the statistical construction and the criteria that support them (Otero 2006, 50).

Camargo (2018) conceives of the first general census of Brazil as a hybrid experiment, pressured by the tradition of partial counts and estimates as well as by the modernity of census exhaustiveness aimed at monitoring the population. The 1872 census was, for Camargo, an administrative instrument in the transition between slave and free labour. Considering the fundamental division that this census established between free and slaved individuals, Camargo concludes that the main contribution of the census was to elaborate and order the state categories that would guarantee the orderly transition to the world of free labour, according to the expectations of reality shared by different elites. Consequently, for the author, the census had modern objectives and adopted the conventions recommended by international statistical organisations, but operated with principles of valuation and classification derived from an archaic or traditional conception of the population.

Finally, in relation to the operations of statistical construction of the nation, the literature shows another dimension of the ends pursued by censuses: their external political use. Several authors agree that censuses allowed for the "diffusion, both internally and internationally, of abstract images that characterised the nation" (Otero 2006, 378, authors' translation). Some authors state that during those years, the financing of costly state operations such as censuses are explained by propaganda purposes (Camargo 2009). Indeed, censuses and statistical yearbooks, as well as almanacs and repertoires, became propaganda tools to attract immigrants, mainly from Europe. The photo or snapshot of the nation provided by the encyclopaedic editions of census works constituted the image that many Latin American countries wanted to portray to the rest of the world.

Statistics: From International to Domestic Circulation

Much of the historiography shares the premise that in Latin America the development of nineteenth-century national censuses and statistics was heavily influenced by two foreign models that were disseminated in the region. On the one hand, the model of the statistical practices sanctioned as valid in the International Statistical Congresses held in Europe. On the other hand, the implicit models regarding what a modern state should be shared by political elites. Latin American states needed to feel that they were members of the community of 'civilised nations', which explains part of their interest in taking censuses. Their ability to carry out these costly enquiries signalled that they pursued the requirements of modern states (Loveman 2014, 304). This thesis rests not only on the emphasis the historiography places on the propagandistic aims of nineteenth-century censuses. It is also based on the idea that statistical yearbooks, once incorporated in global networks of exchange of statistical publications, also had a political function in relation to the external recognition of the states that were being built (Estefane 2019).

Within the historiography that studies Latin American statistics, the role of international organisations as a frame of reference for the national experiences is not completely overlooked. The history of Latin American statistics cannot ignore the influence of foreign specialists. Alexandre

Moreau de Jonnès, for example, is an undisputed reference in the history of the region's statistics, especially considering the national experiences of Brazil and Chile. International missions of experts, moreover, had a relevant role in the region, for example, in promoting projects to centralise official statistics in Colombia (Estrada Orrego 2017). However, the direct external influence of statistical internationalism in the region is questioned.

Some studies highlight the complexities of the process of reception of guidelines and conventions emanating from the International Congresses of Statistics. Other interpretations underline the margins of autonomy with which statistics were managed by the individuals in charge of collecting data and the adaptation of instruments and categories to the local realities. Distancing itself from international recommendations, the 1912 Colombian census incorporated racial categories based on colonial classifications (Prieto 2005). In 1938 national specialists ignored the committee of experts from the League of Nations because they considered the differentiation between inactive and unemployed that they recommended as unsuitable for the Colombian reality (Estrada Orrego 2017). Other works highlight the obstacles with which the attempt to assimilate the Bertillon system of classification of illnesses and causes of death in Chilean statistics stumbled upon (Chávez Zúñiga and Soto-Lara 2019). Analysing the statistical discourse of the DGE-B, Camargo (2015) finds that its internal dynamics were more decisive than the genuine effort of the elites to keep up with the standards of statistical internationalism. Although in most cases these are specific points within broader studies rather than detailed works on the circulation and reception of ideas, they account for the awareness and relevance Latin American historiography places on the unequal power relationship that exists in the production and circulation of statistical knowledge.

Only a couple of works focus specifically on understanding the processes of dissemination and reception of statistical concepts and techniques, from the perspective of the history of the circulation of ideas, such as González Bollo (2007b) for Argentina and Victoria Estrada Orrego (2015) for Colombia. Lanata-Briones (2016) enquires on the relationship between, on the one hand, the International Labour Organisation and its statistical conferences and the Argentine cost-of-living index and the men behind its estimation, on the other. However, substantial further research is needed to ascertain the place of Latin American statisticians in the different waves of the globalisation of figures. Future research should also focus on how concepts, classification schemes, and statistical tools that originated in central spots of statistical internationalism were reappropriated by Latin American statisticians.

There is a need to analyse the trans-nationalised world of statistics and the channels for the circulation of ideas that travelled not only across the Atlantic, but also within Latin America. The lack of cross-national histories about the implementation of statistics in Latin American countries is striking. How was the contact between Latin American statisticians at the end of the nineteenth and first half of the twentieth century? What would have been the density of these links? Did they remain or change over time? These questions should be researched considering the possibilities that existed for those connections and exchanges that included the geographical proximity of these countries and the advantage for communication as much of the region shares the same language. Moreover, further studies should examine the influence of the movement of people due to scholarly congresses and conferences as well as the birth of the Latin-Americanism, of Hispanic-Americanist projects, and of the Pan-American movement promoted by the United States. The cultural history of the circulation of ideas, concepts, and statistical instruments within Latin America is undoubtedly pending.

Regarding the national circulation of statistics, several studies explore the diffusion of official numbers within the local written press. Some works aim to develop an interpretation of the role of statistics in the constitution of public opinion among the enlightened elite of Mexico City in the nineteenth century (Medeles Hernández 2012) or understand the dominant place of quantification in the Chilean public sphere to explain the tradition behind the intensive use of numbers in the public debate towards the last decades of the twentieth and the beginning of the twenty-first century (Márquez Arellano 2010). Other studies conceive of the printed press as a means to comprehend the general public's reactions to the routine tasks of counting that the Argentine state carried out in a context of the expansion of official statistical inquiries between 1890 and 1930 (Daniel and González Bollo 2010). This paper identifies diverse reactions within Argentine society, ranging from the social recognition of the main individuals behind local statistics to the questioning of official figures through criticism, mistrust, and resistance from different social groups to the statistical practices of the state.

Statistics, Scientific Cultures, and Technical Instruments

The contributions of the socio-political history of Latin American statistics expand when considering the study of broader statistical communities, or local scientific societies, on the one hand, and of the analysis of measurement instruments, on the other hand. Following this first line of work, Mayer Celis (1999, 23) recreates the position of statistics within the national scientific culture of republican Mexico during the first half of the nineteenth century, in the search for an applied science, useful to society and to the purposes of the state itself. Mayer Celis shows the statistical interest that existed in formal academic societies, created from the 1830s and made up of true networks of scientists, soldiers, professors, and those interested in academic life, while she ponders what statistics offered to the development of a national imaginary.

During the nineteenth century, the desire to count increased given the successive epidemic waves that hit several Latin American countries. Hence, the links between quantification practices and medical knowledge became attractive to Latin American scholars. The key role of sanitary demographers in Brazil is highlighted in the institutionalisation of Brazilian statistics. Men trained in epidemiology and public health, devoted to solving the serious problems of unhealthy conditions in large cities like Rio de Janeiro, participated in controversies with the official numbers until they reached the DGE-B in 1907 thanks to José Luiz Sayão de Bulhões Carvalho, one of its representatives (Santos 2006; Camargo 2007). In Mexico, several studies examine the link between physicians and the development of statistical practices (Agostoni and Ríos Molina 2010; Cházaro 2005, 2010, 2018). Adopting a broad notion of quantification, Cházaro (2008) analyses the instruments for measuring human bodies that were essential in the medical diagnoses of the nineteenth century, and how these measures were translated into moral hierarchies between genders and races in Mexico. In other works, Cházaro (2010, 2016) explores the encounter between medical research and population statistics in its relationship with the emergence of life insurance in the nineteenth century that made life an insurable asset in the market. In this way, she opens a peephole through which to observe the profound transformations that the notion of human life—as a political good—underwent in Mexico in that period.

The links between statistics, the medical community, and public health are barely explored within the Chilean (Chávez Zúñiga and Soto-Lara

2019) and Argentine (Di Liscia 2009; Daniel 2012a) historiography. Generally, works on Argentina are more prone or oriented to scrutinise socio-labour statistics or the numbers and instruments that contributed to delineating the contours of the so-called 'social question' (González Bollo 2004, 2012b; Daniel 2011a, 2013; Daniel and Vommaro 2017; Lanata-Briones 2021). However, María Silvia Di Liscia (2009) studies the chain for the elaboration of demographic information in its relationship with state healthcare institutions, paying special attention to the differential process of state establishment in the Argentine national territories during the late nineteenth and early twentieth centuries. In the Chilean case, Pablo Chávez Zúñiga and José Soto-Lara (2019) trace the creation of medical-demographic records in Santiago to understand the rise of childhood medical statistics as well as to comprehend the reasons for its emergence and the difficulties it faced. The authors hypothesise that the way of understanding diseases and the causes of infant mortality were modified by the quantification and accumulation of bureaucratic and medical knowledge.

The practices of quantifying crime, criminality, and suicides in Latin America—moral statistics as they were called in nineteenth-century Europe—are also examined through the lens of the socio-political history of statistics. In the scenario of the nascent Mexican republic, Mayer Celis (1999) reviews the proliferation of these figures, the interpretative readings of the "wise men" of the time, and the constitution of debates in newspapers around the deviations from the norm. In a similar sense, Claudia Daniel (2011d) analyses the role of these numbers during the conservative order in Argentina (1880–1916), exploring the statistical objectification tasks of the police of the City of Buenos Aires, and the discussions in which crime and crime-related figures were involved.

The configuration of education as an object of quantification is also an area of interest. The Chilean census of 1854 is analysed as a manifestation of state will power to measure the social bases from which to build the national public education system (Ponce de León et al. 2010). In Chile in the mid-nineteenth century measuring literacy was a way of establishing and evaluating the educational task that still lay ahead for the state. Unlike this specific approach, the socio-historical studies of Natália de Lacerda Gil (2007) on the crossroads between statistics and education in Brazil involve a broader temporality—from the 1870s to the 1940s—while offering a perspective focused on the official figures of Brazilian primary schooling and education. Her work also raises questions that concern the processes by which the legitimacy of educational statistics was developed in Brazil and the way in which these numbers contributed to the creation of shared representations about primary school.

Concerning the links between statistical language and the activities of some professional groups, Jorge Pantaleón (2009) analyses the creation of a "national statistical space" in Argentina between 1918 and 1952 favoured by the cross contribution of disciplines such as statistics, economics, and state planning that were then established as state knowledge. Jimena Caravaca and Claudia Daniel (2019) explore for that same period other aspects of this relationship, especially at the level of professional training and university education. Lanata-Briones (2021) examines the first two estimates of the Argentine cost-of-living index that were released almost in parallel to its United States, French, British, and German counterparts. Her work is an exercise in historical contextualisation and critical methodological reflection regarding one of the crucial statistical tools used by professional economists that would soon be incorporated into the professional culture of the discipline. Works like this remind us of the relevance of articulating the social history of statistical agencies with the technical history of cognitive schemes that accounts for scientific models as well as technical and conceptual tools.

Compared to demographic statistics, socio-economic numbers are relatively less examined. This trend is not exclusive to Latin American statistics, but to statistics more broadly (Tooze 2001, 3). However, in recent years there has been a movement towards exploring them at a regional level. Lanata-Briones (2021) concludes that for cost-of-living indices to hold as stable social and political artefact during the first half of the twentieth century they need to have a role within the political economy; specifically, there had to be a connection between the index and industrial relations. Lanata-Briones and González Bollo (2017) analyse Argentine national income estimates and find that from the mid-1930s and throughout the 1940s two groups of experts were formed within the Argentine state that had responsibilities in the design and implementation of economic programmes and policies. Based on their specific use of statistics, each group had its own vision of the national economy.

Following the assessment of economic statistics, Lanata-Briones (2016) proposes a three-step methodology for the analysis of statistics that can be applied to any estimate. The methodology implies an in-depth analysis of the way in which an estimate is produced and, if needed, it provides a basis to generate a new series. The relevance of each phase depends on the characteristics of the number that is being analysed. Following it, Lanata-Briones (2020a, 2020b) examines the first two estimates of the Argentine cost-of-living index and produces novel estimates of it. Agustina Rayes (2013) generates a specific methodology that reconstructs the complete geographical distribution of Argentine exports for the period 1895–1913.

Recent Trends of Research

To conclude, it is worth highlighting the emerging lines of research that explore other aspects of the socio-political history of Latin American statistics.

Linked to the incipient focus on socio-economic statistics, there is also an emerging trend to focus on the period after the Great Depression. The examination of the dispute the Argentine consumer price index was involved in between 2007 and 2015 suggests that the index became the object of political struggle rather than a methodological controversy (Daniel and Lanata-Briones 2019). Through the analysis of both the Argentine consumer price index between 2007 and 2015 and the attempts to lower Brazilian debt and deficit figures between 2012 and 2015, Roberto Aragão and Lucas Linsi (2022) suggest that macroeconomic indicators are much more ambiguous than it is commonly acknowledged. Therefore, the line between accurate and manipulated data is fuzzier than typical narratives about manipulation suggest. There are also many works that are now well-advanced MSc or PhD projects. The topics being researched include the analysis of the Brazilian consumer price index, the examination of the first estimates of the Mexican cost-of-living index, and the assessment of the notion of economically active population using Argentina as a case study. With the increasing questioning experienced by indicators like the gross domestic product (Coyle 2014; Lepenies 2016), we expect a growing number of works that examine in detail Latin American economic statistics produced during the twentieth and twenty-first centuries.

Inspired by the bumpy trajectory of the Argentine consumer price index between 2007 and 2015, Celia Lury and Ana Gross (2014) are among the first to introduce the analysis of the processes of statactivism to Latin America. Statactivism encompasses a wide variety of practices that place statistics at the heart of political emancipation, thus becoming a form of opposition to neoliberal modes of exercising power (Bruno et al. 2014). Lury and Gross examine the proliferation of alternative measures of inflation in Argentina and suggest that specific configurations of connectivity in the calculative infrastructure enact a space of possibility for statactivism. The COVID-19 crisis and the need to account for the number of femicides in Latin America have opened the door for more studies of this type within the region. For example, Alexandre Camargo, Eugênia Motta, and Victor Mourão (2021) examine the generation of COVID-19 statistics in Brazil, while Natalia Romero Marchesini (2021) analyses the production of official numbers on femicides and travesticides in Argentina.

The historiography on the socio-political history of statistics must also enhance its knowledge of the Caribbean. Several works on the countries and territories of this part of the world comprise the colonial years and the quantification practices followed by the Spanish (Román Antequera 2019) and French empires. Mathieu Aguilera (2017) examines the 1827 and 1841 Cuban colonial population counts and shows how colonial authorities under the control of the army produced demographic data to facilitate successive territorial reforms of the colonial administration. Fanny Malègue (2018) argues that in the aftermath of the Seven Years' War colonial censuses in the French Caribbean were conducted not simply to enumerate the population. They also played a major role in the administration of these slave societies. A few works exist on Puerto Rico. While some examine the relationship between censuses in the United States and in the island (Loveman 2007b), others explore the construction of racial classifications and race in that country (Loveman and Muniz 2007).

* * *

There is substantial research on the socio-political history of Latin American statistics. The four organising nuclei proposed in this bibliographical essay account for the robustness and originality of the regional scholarly production in recent decades. Even though it may still be early to talk about the existence of a properly established field of study, this inventory offers strong arguments in favour of the purpose of this compilation, which aims to situate Latin America as a region that cannot be overlooked when analysing the global history of modern statistics. This has been precisely the objective of the two generations of scholars whose works converge here, the one that enriched the study of statistics by informing it under the influence of social constructivism, and the one that is currently making efforts to enlarge the scope of this turn by considering the questions that the twenty-first century is posing and will continue to pose onto the socio-political history of numbers. As new generations of scholars specialised in Latin America continue to learn about this trajectory—identifying its successes and acknowledging its limitations—there are enough reasons to hope that in the coming decades there will be a more complete vision of the contributions of Latin America both to the global history of knowledge production and to the history of statistics.

References

Excluding the Works Concerning the Socio-Political History of Latin American Statistics

Adelman, Jeremy. 2017. "What Is Global History Now?" *AEON*, 2 March. On-line: https://aeon.co/essays/is-global-history-still-possible-or-has-it-had-its-moment. Accessed 29 April 2021.

Anderson, Benedict. 1983. *Imagined Communities. Reflections on the Origin and Spread of Nationalism*. London: Verso.

Appadurai, Arjun. 2020. "Globalization and the Rush to History." *Global Perspectives* 1 (1): 11656. https://doi.org/10.1525/001c.11656.

Arrext, Carmen, Rolando Mellafe, and Jorge Somoza. 1983. *Demografía histórica en América Latina. Fuentes y métodos*. San José: CELADE.

Balán, Jorge. 2013. "Latin American Higher Education Systems in a Historical and Comparative Perspective." In *Latin America's New Knowledge Economy: Higher Education, Government, and International Collaboration*, edited by Jorge Balán. New York: Institute of International Education.

Brown, Matthew. 2015. "The Global History of Latin America." *Journal of Global History* 10 (3): 365–386. https://doi.org/10.1017/s1740022815000182.

Brunner, José J. 1990. *Educación superior en América Latina: cambios y desafíos*. Santiago: Fondo de Cultura Económica.

Bruno, Isabelle, Emmanuel Didier, and Julien Prévieux, eds. 2014. *Statactivisme: Comment lutter avec les nombres*. Paris: Zones.

Cavarozzi, Marcelo. 1996. *Autoritarismo y democracia (1955–1996). La transición del estado al mercado en la Argentina*. Buenos Aires: Ariel.

Centeno, Miguel A., and Agustin E. Ferraro, eds. 2013. *State and Nation Making in Latin America and Spain: Republics of the Possible*. Cambridge and New York: Cambridge University Press. https://doi.org/10.1017/cbo9781139342667.

CEPAL. Comisión Económica para América Latina y el Caribe. 2009. "Cincuentenario del CELADE: notas sobre su historia y celebración." *Notas de población* 86: 7–17.

Chernilo, Daniel. 2020. "Beyond the Nation? Or Back to It? Current Trends in the Sociology of Nations and Nationalism." *Sociology* 54 (6): 1072–1087. https://doi.org/10.1177/0038038520949831.

Conrad, Sebastian. 2017. *What Is Global History?* Princeton: Princeton University Press. https://doi.org/10.2307/j.ctvc779r7.

Corcuff, Philippe. 2013. *Las nuevas sociologías. Principales corrientes y debates, 1980–2010*. Buenos Aires. Siglo Veintiuno, 2013.

Coyle, Diane. 2014. *GDP. A Brief but Affectionate History*. Princeton: Princeton University Press. https://doi.org/10.1515/9781400873630.

De Lima Grecco, Gabriela and Sven Schuster. 2020. "Decolonizing Global History? A Latin American Perspective." *Journal of World History* 31 (2): 425–446. https://doi.org/10.1353/jwh.2020.0024.

del Castillo, Lina. 2018. *Crafting a Republic for the World: Scientific, Geographic, and Historiographic Inventions of Colombia*. Lincoln and London: University of Nebraska Press. https://doi.org/10.2307/j.ctvgd1ts.

Devés, Eduardo. 2003. *El pensamiento latinoamericano en el siglo XX. Tomo II*. Buenos Aires: Editorial Biblos, Centro de Investigaciones Diego Barros Arana.

Drayton, Richard. 2020. "Response to Arjun Appadurai's 'Globalization and the Rush to History.'" *Global Perspectives* 1 (1): 14275. https://doi.org/10.1525/gp.2020.14275.

Dunkerley, James, ed. 2002. *Studies in the Formation of the Nation-State in Latin America*. London: Institute of Latin American Studies.

Escalante, Fernando. 2019. *Historia mínima del neoliberalismo*. Ciudad de México: El Colegio de México.

Hunt, Lynn. 2015. *Writing History in the Global Era*. New York: W. W. Norton.

Joseph, Gilbert M., and Daniel Nugent. 1994. *Everyday Forms of State Formation: Revolution and the Negotiation of Rule in Modern Mexico*. Durham and London: Duke University Press.

Lepenies, Phillip. 2016. *The Power of a Single Number: A Political History of GDP*. New York, NY: Columbia University Press. https://doi.org/10.7312/columbia/9780231175104.001.0001.

López-Alves, Fernando. 2000. *State Formation and Democracy in Latin America, 1810–1900*. Durham and London: Duke University Press. https://doi.org/10.1215/9780822399636.

Lozoya, Ivette. 2020. *Intelectuales y revolución. Científicos sociales latinoamericanos en el MIR chileno (1965–1973)*. Santiago: Ariadna Ediciones.

Miller, Nicola. 2020. *Republics of Knowledge: Nations of the Future in Latin America*. New Jersey: Princeton University Press. https://doi.org/10.1515/9780691185835.

Pettinà, Vanni. 2018. *Historia mínima de la Guerra Fría en América Latina*. Ciudad de México: El Colegio de México. https://doi.org/10.2307/j.ctv8bt0xr.

Randeraad, Nico. 2010. *States and Statistics in the Nineteenth Century: Europe by Numbers*. Manchester: Manchester University Press. https://doi.org/10.7765/9781526147530.

Recchini de Lattes, Zulma. 1982. "Female Workers Undercounted: The Case of Latin American and Caribbean Censuses." The Population Council, Regional Office for Latin America and the Caribbean. Working Paper N° 12. Mexico City.

Recchini de Lattes, Zulma, and Catalina Wainerman. 1979. *Data from Censuses and Household Surveys for the Analysis of Female Labour in Latin America and the Caribbean: Appraisal of Deficiencies and Recommendations for Dealing with Them*. Santiago: CEPAL.

Ruiz Encina, Carlos. 2019. *La política en el neoliberalismo. Experiencias latinoamericanas*. Santiago: LOM Ediciones.

Sánchez-Albornoz, Nicolás. 1977. *La población de América Latina. Desde los tiempos precolombinos al año 2020*. Madrid: Alianza Universidad.

Tooze, J. Adam. 2001. *Statistics and the German State, 1900–1945: The Making of Modern Economic Knowledge*. New York: Cambridge University Press.

Urrutia, Miguel, and Mario Arrubla, eds. 1970. *Compendio de estadísticas históricas de Colombia*. Bogotá: Universidad Nacional de Colombia.

van Ittersum, Martine, and Jaap Jacobs. 2012. "Are We All Global Historians Now? An Interview with David Armitage." *Itinerario* 36 (2): 7–28. https://doi.org/10.1017/s0165115312000551.

Vessuri, Hebe M. C. 2006. "Academic Science in Twentieth-Century Latin America." In *Science in Latin America. A History*, edited by Juan José Saldaña. Austin: University of Texas Press. https://doi.org/10.7560/712713.

Ward, Michael. 2004. *Quantifying the World. UN Ideas and Statistics*. Bloomington: Indiana University Press.

Yáñez, Juan Carlos. 2008. *La intervención social en Chile y el nacimiento de la sociedad salarial, 1907–1932*. Santiago: RIL Editores.

Socio-Political History of Latin American Statistics

Agostoni, Claudia, and Carlos Andrés Ríos Molina. 2010. *Las estadísticas de salud en México: ideas, actores e instituciones, 1810–2010*. México, D.F.: UNAM, Instituto de Investigaciones Históricas / Secretaría de Salud, Dirección General de Información en Salud.

Aguilera, Mathieu. 2017. "L'ingénieur, les capitaines et les planteurs. Le recensement de la *Siempre Fiel Isla de Cuba* (1825–1842): entre savoirs locaux et préoccupations impériales." *Histoire & Mesure* 32 (1): 9–52. https://doi.org/10.4000/histoiremesure.5747.

Alves, Marilda D. 1988. *O desenvolvimento do Sistema Estatístico Nacional*. Rio de Janeiro: Fundação Instituto Brasileiro de Geografia e Estatística. https://biblioteca.ibge.gov.br/visualizacao/livros/liv19387.pdf. Accessed 28 September 2020.

Angosto Ferrández, Luis Fernando, and Sabine Kradolfer, eds. 2012. *Everlasting Countdowns: Race, Ethnicity and National Censuses in Latin American States*. Newcastle upon Tyne: Cambridge Scholars Publishing.

Aragão, Roberto, and Lukas Linsi. 2022. "Many Shades of Wrong: What Governments do When They Manipulate Statistics." *Review of International Political Economy* 29 (1): 88–113. https://doi.org/10.1080/09692290.2020.1769704.

Arrioja Díaz Viruell, Luis Alberto, ed. 2016. *Registrar e imaginar la nación: la estadística durante la primera mitad del siglo XIX*. Zamora: El Colegio de Michoacán.

Augustine-Adams, Kif. 2015. "Hacer a México: la nacionalidad, los chinos y el censo de población de 1930." In *Inmigración y racismo. Contribuciones a la historia de los extranjeros en México*, edited by Pablo Yankelevich. México, D.F.: El Colegio de México.

Azevedo, Aloysio Villela de. 1990. *Os recenseamentos no Brasil*. Rio de Janeiro: IBGE. https://biblioteca.ibge.gov.br/visualizacao/livros/liv84381.pdf. Accessed 28 September 2020.

Barickman, Bert J. 2003. "Reading the 1835 Parish Censuses from Bahia: Citizenship, Kinship, Slavery and Household in Early Nineteenth-Century Brazil." *The Americas* 59 (3): 287–323. https://doi.org/10.1353/tam.2003.0002.

Braun Llona, Juan, Matías Braun Llona, Ignacio Briones, José Díaz, Rolf Lüders, and Gert Wagner. 2000. "Economía chilena 1810–1995: estadísticas históricas." *Documento de Trabajo* 187, Instituto de Economía, Pontificia Universidad Católica de Chile. http://www.memoriachilena.gob.cl/602/w3-article-9044.html. Accessed 28 September 2020.

Bustamante, Jesús, Laura Giraudo, and Leticia Mayer. 2014. *La novedad estadística. Cuantificar, cualificar y transformar las poblaciones en Europa y América Latina, siglos XIX y XX*. Madrid: Ediciones Polifemo.

Cahill, David. 1994. "Colour by Numbers: Racial and Ethnic Categories in the Viceroyalty of Peru, 1532–1824." *Journal of Latin American Studies* 26 (2): 325–346. https://doi.org/10.1017/S0022216X00016242.

Camargo, Alexandre de Paiva Rio. 2007. "Números para o progresso: um panorama da actividade estatística na Primeira República." In *Bulhões Carvalho, Um Médico Cuidando da Estatística Brasileira*, edited by IBGE, Centro de Documentação e Disseminação de Informações. Rio de Janeiro: IBGE. https://biblioteca.ibge.gov.br/visualizacao/monografias/GEBIS%

20-%20RJ/ColecaoMemoriaInstitucional/11-Um%20Medico%20Cuidando%20da%20Estatistica-Bulhoes%20Carvalho.pdf. Accessed 28 September 2020.
———. 2008. "Municipalismo e ruralismo: o IBGE e a organização nacional na era Vargas." In *O IBGE na história do municipalismo e sua atuação nos municípios*, edited by Nelson de Castro Senra. Rio de Janeiro: IBGE. https://biblioteca.ibge.gov.br/visualizacao/livros/liv38889.pdf. Accessed 12 May 2021.
———. 2009. "Mensuração racial e campo estatístico nos censos brasileiros (1872–1940): uma abordagem convergente." *Boletim do Museu Paraense Emílio Goeldi*. Ciências Humanas 4: 361–385. https://doi.org/10.1590/S1981-81222009000300002.
———. 2010. "Povoar o hinterland: o ensino rural como fronteira entre estatística e educação na trajetória de Teixeira de Freitas." *Revista Brasileira de História da Educação* 10 (2): 97–132. https://periodicos.uem.br/ojs/index.php/rbhe/article/view/38531. Accessed 12 May 2021.
———. 2011. "A demografia no discurso médico-higienista: um estudo baseado no Brazil-Médico (1887–1900)." In *Associação das Américas, as estatísticas públicas como objeto de estudo*, edited by Senra, N. et al., SEI (série Estudos e Pesquisas).
———. 2012. "Demografia sanitária e a emergência de um estilo de raciocínio estatístico na Primeira República." In *Anais do XVIII Encontro Nacional de Estudos Populacionais*, edited by Associaçã Brasileira de Estudos Populacionais. Águas de Lindóia: Associaçã Brasileira de Estudos Populacionais. http://www.abep.org.br/publicacoes/index.php/anais/article/viewFile/2032/1990. Accessed 28 September 2020.
———. 2013. "Estatística e política de população no Brasil (1870–1930): notas de pesquisa." *Estatística e Sociedade* 3: 90–109. https://seer.ufrgs.br/estatisticaesociedade/article/view/45061. Accessed 11 February 2022.
———. 2015. "Dimensões da nação: uma análise do discurso estatístico da diretoria geral de estatística (1872–1930)." *Revista* Brasileira *de Ciencias Sociais* 30: 79–97. https://doi.org/10.17666/308779-95/2015.
———. 2016. "A construçao da medida comum: estatística e política de populaçao no Império e na Primeira República." PhD dissertation, Universidade do Estado do Rio de Janeiro.
———. 2017. "Sur l'émergence d'une politique de précision. Le cas des statistiques brésiliennes." *Histoire & Mesure* 32 (1): 91–119. https://doi.org/10.4000/histoiremesure.5807.
———. 2018. "O censo de 1872 e a utopia estatística do Brasil Imperial." *Revista de História da Unisinos* 22 (3): 414–428. http://revistas.unisinos.br/index.php/historia/article/view/15474. Accessed 8 January 2021.
———. 2021a. "Para uma arqueologia da verdade estatística: reflexões a partir do projeto de reforma educacional de Rui Barbosa (1882)." *Revista Brasileira*

de História e Ciências Sociais 13 (25): 500–529. https://doi.org/10.14295/rbhcs.v13i25.11938.

———. 2021b. "Estatísticas sanitárias e interdependência social na Primeira República." *História, Ciências, Saúde – Manguinhos* 28 (4): 1007–1035. https://doi.org/10.1590/S0104-59702021000400006

Camargo, Alexandre de Paiva Rio, Eugênia de Souza Mello Guimarães Motta, and Victor Luiz Alves Mourão. 2021. "Números emergentes: temporalidade, métrica e estética da pandemia de Covid-19." *Mediações - Revista de Ciências Sociais* 26 (2): 311–332. https://doi.org/10.5433/2176-6665.2021v26n2p311

Camargo Pereira, José Severo de, and Pedro A. Morettin. 1991. "Las estadísticas brasileñas y la enseñanza de la Estadística en Brasil." *Estadística Española* 33 (128): 559–574. https://www.ine.es/ss/Satellite?blobcol=urldata&blobheader=application%2Fpdf&blobheadername1=Content-Disposition&blobheadervalue1=attachment%3B+filename%3D128_8.pdf&blobkey=urldata&blobtable=MungoBlobs&blobwhere=564%2F582%2F128_8%2C0.pdf&ssbinary=true. Accessed 28 September 2020.

Caravaca, Jimena, and Claudia Daniel. 2019. "El nacimiento de una profesión. Economistas y saberes estadísticos en la Argentina entre 1913 y 1935." In *Profissões, burocracias e saberes: perspectivas históricas (Brasil/Argentina/Chile - séculos XIX e XX)*, edited by Ana Paula Korndörfer, Jonas Moreira Vargas, and Marcelo Vianna. Brasil: Paco e Litera Editorial.

Carvajal Bañados, Yuri, and Tuillang Yuing Alfaro. 2013. "Gramáticas de vida: el censo de 1907 y la población como uso de gobierno." *História, Ciências, Saúde-Manguinhos* 20 (4): 1473–1489. https://www.scielo.br/pdf/hcsm/v20n4/0104-5970-hcsm-20-04-01473.pdf. Accessed 28 September 2020.

Carvalho Junior, Cesar Vaz de, Edmundo Sá Figueirôa, Nelson de Castro Senra, and Hernán González Bollo, eds. 2011. *Em associação das Américas, as estatísticas públicas como objeto de estudo*. Salvador: Superintendência de Estudos Econômicos e Sociais da Bahia. http://www.sei.ba.gov.br/images/publicacoes/download/sep/sep_90.zip. Accessed 28 September 2020.

Chávez Zúñiga, Pablo, and José Julián Soto-Lara. 2019. "La primera estadística sanitaria infantil de la provincia de Santiago de Chile (1860–1929)." *Dynamis* 39 (1): 149–174. https://revistaseug.ugr.es/index.php/dynamis/article/view/8670/7831. Accessed 28 September 2020.

Cházaro, Laura. 2000a. "La Ciudad de México ante la sanción de las estadísticas de mediados del siglo XIX; entre los miasmas y la podredumbre de la civilización." In *Instituciones y Ciudad. Ocho estudios históricos sobre la Ciudad de México*, edited by Carlos Illades and Ariel Rodríguez Kuri. México, D.F.: Ediciones Unión.

———. 2000b. "Medir y valorar los cuerpos de una nación. Un ensayo sobre la estadística médica en el siglo XIX." PhD dissertation, UNAM.

———. 2001. "Imágenes de la población mexicana: descripciones, frecuencias y cálculos estadísticos." *Relaciones. Estudios de Historia y Sociedad* 22 (88): 15–48. https://www.redalyc.org/articulo.oa?id=13708802. Accessed 28 September 2020.

———. 2005. "Reproducción y muerte de la población mexicana: Cálculos estadísticos y preceptos higiénicos a fines del siglo diecinueve." In *Discursos, prácticas y sanciones. Ensayos de historia social en una perspectiva comparada (México, Brasil y Argentina, 1850–1950)*, edited by Claudia Agostoni and Elisa Speckman. México, D.F.: UNAM, Instituto de Investigaciones Históricas.

———. 2008. "Regímenes e instrumentos de medición: Las medidas de los cuerpos y del territorio nacional en el siglo XIX en México." *Nuevo Mundo Mundos Nuevos*. Accessed 11 January 2021. https://doi.org/10.4000/nuevomundo.14052

———. 2010. "La vida y su valor a discusión: juicios médicos y cálculos de probabilidades entre médicos mexicanos del siglo XIX." In *Estadisticas Nas Americas. Por uma agenda de estudos históricos comparados*, edited by Nelson de Castro Senra and Alexandre de Paiva Rio Camargo. Rio de Janeiro: IBGE. https://biblioteca.ibge.gov.br/visualizacao/livros/liv44323.pdf. Accessed 28 September 2020.

———. 2012. "Medir la política en la Nueva España de Alexander von Humboldt." In *Economía, ciencia y política. Estudios sobre Alexander von Humboldt a 200 años del Ensayo político sobre el reino de la Nueva España*, edited by José Enrique Covarrubias and Matilde Souto Mantecón. México, D.F.: UNAM, Instituto de Investigaciones Históricas / Instituto Mora.

———. 2016. "¿El valor de la vida y del trabajo? Las compañías de Seguros de vida, México a fines del s. XIX." *Estudios Sociales del Estado* 2 (4): 74–95. http://www.estudiossocialesdelestado.org/index.php/ese/article/view/91. Accessed 28 September 2020.

———. 2018. "Médecins, statistique et recensement: les raisons d'une entente cordiale." *Histoire et Mesure* 33 (2): 33–60. https://doi.org/10.4000/histoiremesure.7970.

Chiaramonti, Gabriella. 2000a. "Buscando al ciudadano 'virtuoso'. El censo peruano de 1876 en el proyecto político de Manuel Pardo." In *Constitucionalismo y orden liberal: América Latina, 1850–1920*, edited by Marcello Carmagnani. Torino: Otto Editore.

———. 2000b. "La ley y las costumbres. Apuntes sobre los registros civiles y los libros parroquiales en el Perú de la segunda mitad del siglo XIX (1857–1879)." *Revista Complutense de Historia de América* 26: 199–232. https://revistas.ucm.es/index.php/RCHA/article/view/RCHA0000110199A. Accessed 28 September 2020.

Clark, A. Kim. 2002. "Race, 'Culture', and Mestizaje: The Statistical Construction of the Ecuadorian Nation, 1930–1950." *Journal of Historical Sociology* 11 (2): 185–211. https://doi.org/10.1111/1467-6443.00060.

Daniel, Claudia. 2009. "El estado argentino y sus estadísticas. El derrotero de un largo proceso de institucionalización (1864–1968)." *Illapa, Revista Latinoamericana de Ciencias Sociales* 5: 151–174. https://es.scribd.com/doc/37441931/Revista-ILLAPA-n%C2%BA-5-agosto-2009. Accessed 28 September 2020.

———. 2010. "Las estadísticas argentinas como objeto de estudio. Ejercicio de formalización de una experiencia de investigación histórica." In *Estatísticas nas Américas. Por uma agenda de estudos históricos comparados*, edited by Nelson de Castro Senra and Alexandre de Paiva Rio Camargo. Rio de Janeiro: IBGE. https://biblioteca.ibge.gov.br/visualizacao/livros/liv44323.pdf. Accessed 28 September 2020.

———. 2011a. "Cuando las cifras componen lo social. Estado, estadísticas y expertos en la construcción histórica de la cuestión social en Argentina (1913–1983)." In *Saber lo que se hace. Expertos y política en Argentina*, edited by Sergio Morresi and Gabriel Vommaro. Buenos Aires: Prometeo / Universidad Nacional de General Sarmiento.

———. 2011b. "L'objectivation des risques, le langage des certitudes. Les statistiques du travail en Argentine pendant la période 1930–1943." *Sociologie et Sociétés* 43 (2): 177–199. https://doi.org/10.7202/1008243ar.

———. 2011c. "Las estadísticas públicas en la Argentina de entreguerras. Agencias, actores y programas de recuento." In *Em associação das Américas, as estatísticas públicas como objeto de estudo*, edited by Cesar Vaz de Carvalho Junior, Edmundo Sá Figueirôa, Nelson de Castro Senra, and Hernán González Bollo. Salvador: Superintendência de Estudos Econômicos e Sociais da Bahia. http://www.sei.ba.gov.br/images/publicacoes/download/sep/sep_90.zip. Accessed 28 September 2020.

———. 2011d. "Medir la moral pública. La cuantificación policial del delito en Buenos Aires, 1880–1910." *Estadística y Sociedad* 1 (1): 149–165. https://seer.ufrgs.br/estatisticaesociedade/article/view/24561/14225. Accessed 28 September 2020.

———. 2012a. "Contar para curar. Estadísticas y comunidad médica en Argentina (1880–1940)." *Historia, Ciencias, Saude-Manguinhos* 19 (1): 89–114. https://doi.org/10.1590/S0104-59702012000100006.

———. 2012b. "Una escuela científica en el Estado. Los estadígrafos oficiales en la Argentina de entreguerras." In *Los Saberes del Estado*, edited by Mariano Plotkin and Eduardo Zimmermann. Buenos Aires: Edhasa.

———. 2013. "De crisis a crisis. La invención de la desocupación en Argentina." *Revista de Indias* 73 (257): 193–218. https://doi.org/10.3989/revindias.2013.008.

———. 2015. "Observar, medir, comparar. La Ciudad de Buenos Aires bajo el lente estadístico de fines del siglo XIX, principios del XX." *REDES* 21 (40): 147–176. http://ridaa.unq.edu.ar/handle/20.500.11807/379. Accessed 28 September 2020.

———. 2016. "Modelos para armar: estadísticas públicas en Argentina y Brasil entre 1930 y 1950." *Revista Estadística Española* 58 (190): 181–204. https://www.ine.es/ss/Satellite?L=0&c=INERevEstad_C&p=1254735226759&pagename=ProductosYServicios%2FPYSLayout&_charset_=UTF-8&cid=1259949703141&submit=Ir. Accessed 28 September 2020.

———. 2017. "Rebuilding a Statistical Apparatus at a Critical Juncture. Argentina and Brazil, from the Great Depression to the Post-World War II Period." *Histoire & Mesure* 32 (1): 161–190. https://doi.org/10.4000/histoiremesure.5797.

———. 2018. "Aparato estadístico, paradigma de la planificación y desarrollismo en Argentina (1955–1970)." *História Unisinos* 22 (4): 620–636. https://doi.org/10.4013/htu.2018.224.09.

Daniel, Claudia, and Cecilia T. Lanata-Briones. 2019. "Battles Over Numbers: The Case of the Argentine Consumer Price Index (2007–2015)." *Economy & Society* 48 (1): 127–151. https://doi.org/10.1080/03085147.2019.1579438.

Daniel, Claudia, and Gabriel Vommaro. 2017. "Poor in Numbers: Contribution to a Social History of Social Statistics in Contemporary Argentina." *Journal of Latin American Studies* 49 (3): 463–487. https://doi.org/10.1017/S0022216X16001899.

Daniel, Claudia, and Hernán González Bollo. 2010. "Las estadísticas oficiales en la prensa escrita porteña (Argentina, 1890–1930)." In *Estatísticas nas Américas. Por uma agenda de estudos históricos comparados*, edited by Nelson de Castro Senra and Alexandre de Paiva Rio Camargo. Rio de Janeiro: IBGE. https://biblioteca.ibge.gov.br/visualizacao/livros/liv44323.pdf. Accessed 28 September 2020.

De la Peña, Sergio, and James Wilkie. 1994. *La estadística económica en México. Los orígenes.* México, D.F.: Siglo XXI Editores.

Di Liscia, María Silvia. 2009. "Cifras y problemas. Las estadísticas y la salud en los Territorios Nacionales (1880–1940)." *Salud Colectiva* 5 (2): 259–278. https://doi.org/10.18294/sc.2009.264.

Ervin, Michael. 2007. "The 1930 Agrarian Census in Mexico: Agronomists, Middle Politics, and the Negotiation of Data Collection." *Hispanic American Historical Review* 87: 537–570. https://doi.org/10.1215/00182168-2007-004.

———. 2009. "Statistics, Maps, and Legibility: Negotiating Nationalism in Post-Revolutionary Mexico." *The Americas* 66 (2): 155–179. https://doi.org/10.1353/tam.0.0165.

Estefane, Andrés. 2004. "'Un alto en el camino para saber cuántos somos…'. Los censos de población y la construcción de lealtades nacionales. Chile, siglo XIX." *Historia* 37 (1): 33–59. https://red.pucp.edu.pe/riel/files/2015/04/Estefane_Censos_XIX.pdf. Accessed 28 September 2020.

———. 2012a. "Burócratas ambulantes. Movilidad y producción de conocimiento estadístico en Chile, 1860–1873." *Revista Enfoques. Ciencia Política y Administración Pública* 10 (17): 123–146. http://www.revistaenfoques.cl/index.php/revista-uno/article/view/68/49. Accessed 28 September 2020.

———. 2012b. "Enumerar lo que se gobierna. La producción del *Anuario Estadístico de la República de Chile*." In *Anuario Estadístico de la República de Chile*, edited by Cámara Chilena de la Construcción, Pontificia Universidad Católica de Chile and Biblioteca Nacional de Chile. Santiago de Chile: Cámara Chilena de la Construcción / UC / Biblioteca Nacional de Chile. http://www.bibliotecanacionaldigital.gob.cl/bnd/632/w3-article-355660.html. Accessed 28 September 2020.

———. 2016. "La institucionalización del sistema estadístico chileno: debates y problemas prácticos (1843–1851)." *Estudios Sociales del Estado* 2 (4): 35–73. http://www.estudiossocialesdelestado.org/index.php/ese/article/view/90. Accessed 28 September 2020.

———. 2017. "Elusive Numbers. State Knowledge and Bureaucratic Organization in Chile (1750–1911)." PhD dissertation, State University of New York at Stony Brook. http://hdl.handle.net/11401/78299. Accessed 28 September 2020.

———. 2019. "Bureaucracy and State Knowledge: On the Production of Statistics in Chile (1750s–1870s)." *Nuevo Mundo Mundos Nuevos*. http://journals.openedition.org/nuevomundo/76841. Accessed 28 September 2020.

Estefane, Andrés, and Elvira López. 2019. "'La primera colección de datos estadísticos de toda la República'. El cuestionario de Claudio Gay de 1841 y la producción de saber estatal en Chile republicano." *Revista de Historia y Geografía UCSH* 41: 49–78. http://ediciones.ucsh.cl/ojs/index.php/RHyG/article/view/2090/1881. Accessed 28 September 2020.

Estrada Orrego, Victoria. 2015. "La valeur des chiffres: La production et les usages des statistiques démographiques et de santé publique en Colombie 1886–1947." PhD dissertation, École des Hautes Études en Sciences Sociales.

———. 2017. "¿Cuántos somos? Una historia de los censos civiles y de la organización de la estadística en Colombia en la primera mitad del siglo XX." *Historia Crítica* 64: 141–160. https://doi.org/10.7440/histcrit64.2017.08

Ferreiro, Osvaldo, and Carlos Prado. 1991. "Historia de la Estadística en Chile." *Estadística Española* 33 (128): 575–586. https://www.ine.es/ss/Satellite?L=0&c=INERevEstad_C&p=1254735226759&pagename=ProductosYServi

cios%2FPYSLayout&_charset_=UTF-8&cid=1259925029848&submit=Ir. Accessed 28 September 2020.

Gil, Natalia de Lacerda. 2007. "A dimensão da educação nacional: um estudo sócio-histórico sobre as estatísticas oficiais da escola brasileira." PhD dissertation, Universidade de São Paulo. https://www.teses.usp.br/teses/disponiveis/48/48134/tde-31052007-112812/en.php. Accessed 14 June 2021.

———. 2020. "Quantification of an Educational System. Numbers in the Social Differentiation of Brazil." In *The International Emergence of Educational Sciences in the Post-World War Two Years. Quantification, Visualization, and Making Kinds of People*, edited by Thomas S. Popkewitz, Daniel Pettersson, Kai-Jung Hsiao. New York: Routledge. https://doi.org/10.4324/9780429060540.

Gómez, Fernando. 1970. "Los censos en Colombia antes de 1905." In *Estudio de estadísticas históricas de Colombia*, edited by Miguel Urrutia and Mario Arrubla. Bogota: Universidad Nacional de Colombia.

González Bollo, Hernán. 1999a. "Ciencias Sociales y Sociografía Estatal. Tras el Estudio de la Familia Obrera Porteña, 1899–1932." *Estudios Sociales* 16 (1): 19–39. https://doi.org/10.14409/es.v16i1.2423.

———. 1999b. "Génesis y funciones de los manuales estadísticos y geográficos (1852–1876)." *Anuario IEHS* 14: 23–42.

———. 2004. "La cuestión obrera en números: la estadística socio-laboral y su impacto en la política y la sociedad, 1895–1943." In *El mosaico argentino. Modelos y representaciones del espacio y de la población, siglos XIX-XX*, edited by Hernán Otero. Buenos Aires: Siglo XXI.

———. 2007a. "La estadística pública y la expansión del estado argentino: una historia social y política de una burocracia especializada, 1869–1947." PhD dissertation, Universidad Torcuato Di Tella. http://repositorio.utdt.edu/handle/utdt/902. Accessed 28 September 2020.

———. 2007b. "Recepción en Argentina de una obsesión demográfica occidental. Baja la natalidad, declina la población, se agranda el desierto, 1926–1943." In *Los lugares del saber*, edited by Ricardo Salvatore. Rosario: Beatriz Viterbo.

———. 2010a. "La elite estadística y el estado argentino (1869–1947)." In *Estatísticas nas Américas. Por uma agenda de estudos históricos comparados*, edited by Nelson de Castro Senra and Alexandre de Paiva Rio Camargo. Rio de Janeiro: IBGE. https://biblioteca.ibge.gov.br/visualizacao/livros/liv44323.pdf. Accessed 28 September 2020.

———. 2010b. "Sobre la amenazante mayoría de dos provincias y una ciudad: los tres primeros censos demográficos y su impacto político en Argentina (1853–1920)." *Estadística Española* 52 (174): 299–319.

———. 2010c. "Transformar la campaña argentina: los expertos de la Dirección de Economía Rural y Estadística del Ministerio de Agricultura, promotores de la cooperación rural (1907–1930)." In *Un Estado con rostro humano. Funcionarios e instituciones estatales en Argentina (1880 hasta la actualidad)*, edited by Ernesto Bohoslavsky and Germán Soprano. Buenos Aires: Prometeo / Universidad Nacional de General Sarmiento.

———. 2012a. *La teodicea estadística de Alejandro E. Bunge*. Buenos Aires: Imago Mundi, UCA.

———. 2012b. "La División Estadística del Departamento Nacional del Trabajo y su transformación en Dirección de Estadística Social del Consejo Nacional de Posguerra (1932–1945)." In *La construcción de las instituciones laborales en la Argentina: el Departamento Nacional del Trabajo, 1907–1943*, edited by Mirta Lobato and Juan Suriano. Buenos Aires: Edhasa.

———. 2014. *La fábrica de las cifras oficiales del Estado argentino (1869–1947)*. Bernal: Universidad Nacional de Quilmes Editorial.

Gootenberg, Paul. 1991. "Population and Ethnicity in Early Republican Peru: Some Revisions." *Latin American Research Review* 26 (3): 109–157. https://www.jstor.org/stable/2503666. Accessed 28 September 2020.

Hutchison, Elizabeth Quay. 2000. "La historia detrás de las cifras: La evolución del censo chileno y la representación del trabajo femenino, 1895–1930." *Historia* 33: 417–434. https://doi.org/10.4067/S0717-71942000003300009.

Lanata-Briones, Cecilia T. 2016. "Constructing Public Statistics: The History of the Argentine Cost of Living Index, 1918–1943." PhD dissertation, London School of Economics and Political Science. http://etheses.lse.ac.uk/3319/. Accessed 9 July 2021.

———. 2020a. "Una nueva estimación del índice del costo de la vida, Argentina 1912–1932." *Boletín del Instituto de Historia Argentina y Americana Dr Emilio Ravignani* 53 (2): 64–92. https://doi.org/10.34096/bol.rav.n53.8006.

———. 2020b. "De-construyendo, construyendo y re-construyendo estadísticas: El índice del costo de la vida, *Argentina* 1933–1937." *Anuario IEHS* 35 (2): 67–99. https://doi.org/10.37894/ai.v35i2.779.

———. 2021. "Constructing Cost of Living Indices: Ideas and Individuals, *Argentina* 1918–1935." *History of Political Economy* 53 (1): 57–87. https://doi.org/10.1215/00182702-8816613.

Lanata-Briones, Cecilia T., and Hernán González Bollo. 2017. "Unwinding the Winding Trajectory: The Divergence Between National Accounts and State Planning, Argentina 1937–1948." *Histoire & Mesure* 32 (1): 161–190. https://doi.org/10.4000/histoiremesure.5790

Lomnitz, Claudio. 1999. "El censo y la transformación de la esfera pública." In *Modernidad Indiana. Nueve ensayos sobre nación y mediación en México*. México, D.F.: Planeta.

Loveman, Mara. 2007a. "Blinded Like a State: The Revolt against Civil Registration in Nineteenth-Century Brazil." *Comparative Studies in Society and History* 49 (1): 5–39. https://doi.org/10.1017/S0010417507000394.

———. 2007b. "The U.S. Census and the Contested Rules of Racial Classification in Early Twentieth-Century Puerto Rico." *Caribbean Studies* 35 (2): 79–114. https://www.redalyc.org/pdf/392/39215017004.pdf. Accessed 28 September 2020.

———. 2007c. "Whiteness in Latin America: Measurement and Meaning in National Censuses (1850–1950)." *Journal de la Société des Américanistes* 95 (2): 207–234. https://doi.org/10.4000/jsa.11085.

———. 2009. "The Race to Progress: Census Taking and Nation Making in Brazil (1870–1920)." *Hispanic American Historical Review* 89 (3): 435–470. https://doi.org/10.1215/00182168-2009-002.

———. 2013. "Census Taking and Nation Making in Nineteenth-Century Latin America." In *State and Nation Making in Latin America and Spain. Republics of the Possible*, edited by Miguel Ángel Centeno and Agustín Ferraro. New York: Cambridge University Press. https://doi.org/10.1017/CBO9781139342667.

———. 2014. *National Colors. Racial Classification and the State in Latin America*. New York: Oxford University Press. https://doi.org/10.1093/acprof:oso/9780199337354.001.0001.

Loveman, Mara, and Jeronimo O. Muniz. 2007. "How Puerto Rico Became White: Boundary Dynamics and Intercensus Racial Reclassification." *American Sociological Review* 72 (6): 915–939. https://doi.org/10.1177/000312240707200604.

Loveman, Mara, Jeronimo O. Muniz, and Stanley R. Bailey. 2012. "Brazil in Black and White? Race Categories, the Census and the Study of Inequality." *Ethnic and Racial Studies* 35 (8): 1466–1483. https://doi.org/10.1080/01419870.2011.607503.

Lozano, María. 1991. "La Sociedad Mexicana de Geografía y Estadística (1833)-1867). Un Estudio de caso: La Estadística." M.A. thesis, Universidad Nacional Autonoma de Mexico, Mexico.

Lury, Celia, and Ana Gross. 2014. "The Downs and Ups of the Consumer Price Index in Argentina." *Partecipazione e Conflitto* 7(2): 258–277. https://doi.org/10.1285/i20356609v7i2p258.

Mamalakis, Markos. 1978. "Historical Statistics of Chile. An Introduction." *Latin American Research Review* 13 (2): 127–137. https://www.jstor.org/stable/2502523. Accessed 28 September 2020.

Marcilio, Maria L. 1977. "Levantamentos Censitários da fase proto-estatística do Brasil." *Anais de Historia* 9: 63–75.

Márquez Arellano, Rodrigo. 2010. "La "medida" de lo posible. Cuantificación y esfera pública en Chile." PhD dissertation, Universiteit Leiden. https://

openaccess.leidenuniv.nl/bitstream/handle/1887/15334/dissertatie%20marquez.pdf?sequence=1. Accessed 28 September 2020.

Malègue, Fanny. 2018. "L'empire en tableaux. Recenser et gouverner les colonies antillaises après la guerre de Sept Ans." *Histoire et Mesure* 33 (2): 93–114. https://doi.org/10.4000/histoiremesure.8038.

Mayer Celis, Leticia. 1999. *Entre el infierno de una realidad y el cielo de un imaginario: Estadística y comunidad científica en el México de la primera mitad del siglo XIX*. México, D.F.: El Colegio de México, Centro de Estudios Históricos. https://doi.org/10.2307/j.ctvhn0d1g.

———. 2000. "La ley de los grandes números y la cultura liberal en México, 1856–1885." In *Institucionalismo y orden liberal. América Latina, 1850–1920*, edited by Marcello Carmagnani. Torino: Otto Editore.

———. 2003. *La tan buscada modernidad científica. Boletín del Instituto Nacional de Geografía y Estadística de 1839*. Mexico: IIMAS-UNAM.

———. 2015. *Rutas de incertidumbre. Ideas alternativas sobre la génesis de la probabilidad, siglos XVI y XVII*. México, D.F.: Fondo de Cultura Económica.

Medeles Hernández, Ana María. 2011. "El pensamiento estadístico, un instrumento medición en México en el siglo XIX." In *Em associação das Américas, as estatísticas públicas como objeto de estudo*, edited by Cesar Vaz de Carvalho Junior, Edmundo Sá Figueirôa, Nelson de Castro Senra, and Hernán González Bollo. Salvador: Superintendência de Estudos Econômicos e Sociais da Bahia. http://www.sei.ba.gov.br/images/publicacoes/download/sep/sep_90.zip. Accessed 28 September 2020.

———. 2011. "Medición y población en México a finales del siglo XIX: estadísticas electorales." M.A. Thesis, Universidad Nacional Autónoma de México.

———. 2012. "Las Estadísticas Públicas En *El Monitor Republicano*: Administrar Con Números." *Estatística e Sociedade* 2: 112–127. https://seer.ufrgs.br/estatisticaesociedade/article/view/36559/23790. Accessed 28 September 2020.

———. 2016. "Documento histórico. La ley del 26 de mayo de 1882 que constituyó la Dirección General de Estadística." *Estadística y Sociedad* 4: 153–159. https://seer.ufrgs.br/estatisticaesociedade/article/view/64424. Accessed 27 May 2021.

Mentz, Raúl. 1991. "Sobre la historia de la estadística oficial argentina." *Estadística Española* 33 (128): 501–532. https://www.ine.es/ss/Satellite?L=0&c=INERevEstad_C&p=1254735226759&pagename=ProductosYServicios%2FPYSLayout&_charset_=UTF-8&cid=1259925029848&submit=Ir. Accessed 28 September 2020.

Mentz, Raúl, and Víctor Yohai. 1991. "Sobre la historia de la enseñanza de la estadística en las universidades argentinas." *Estadística Española* 33 (128): 533–558. https://www.ine.es/ss/Satellite?L=0&c=INERevEstad_C&p=125

4735226759&pagename=ProductosYServicios%2FPYSLayout&_charset_=UTF-8&cid=1259925029848&submit=Ir. Accessed 28 September 2020.

Miño Grijalva, Wilson. 2015. *Una mirada histórica a la estadística del Ecuador*. Quito: Instituto Nacional de Estadísticas y Censos. https://www.ecuadorencifras.gob.ec/documentos/web-inec/Bibliotecas/Libros/INEC_Historia_Censos.pdf. Accessed 28 September 2020.

Necochea López, Raúl. 2010. "Demographic Knowledge and Nation-Building: The Peruvian Census of 1940." *Ber Wiss* 33 (3): 280–296. https://doi.org/10.1002/bewi.201001471.

Olivier, Martin. 2001. "Da estatística política à sociologia estatística. Desenvolvimento e transformações da análise estatística da sociedade (séculos XVII–XIX)." *Revista Brasileira de História* 21 (41): 13–34. https://doi.org/10.1590/S0102-01882001000200002.

Otero, Hernán. 1997–1998. "Estadística censal y construcción de la Nación. El caso argentino, 1869–1914." *Boletín del Instituto de Historia Argentina y Americana "Dr. Emilio Ravignani"* 16–17: 123–149. http://ravignanidigital.com.ar/_bol_ravig/n16_17/n1617a05.pdf. Accessed 28 September 2020.

———. 1999. "Demografía política e ideología estadística en la estadística censal argentina, 1869–1914." *Anuario del Instituto de Estudios Histórico-Sociales* 14: 43–70. http://anuarioiehs.unicen.edu.ar/Files/1999/003%20-%20Otero,%20Hernan%20-%20Demografia%20politica%20e%20ideolog%C3%ADa%20estad%C3%ADstica%20en%20la%20estad%C3%ADstica%20censal%20argentina%201869-1914.pdf. Accessed 28 September 2020.

———. 2004. "Crítica de la razón estadística. Ensayo de formalización teórico-metodológica del paradigma censal de la Argentina moderna." In *El mosaico argentino: modelos y representaciones del espacio y de la población. Siglos XIX y XX*, edited by Hernán Otero. Buenos Aires: Siglo XXI.

———. 2006. *Estadística y Nación. Una historia conceptual del pensamiento censal de la Argentina moderna, 1869–1914*. Buenos Aires: Prometeo.

———. 2007. "Censos antiguos: 1869, 1895, 1914, 1947." In *Población y bienestar en la Argentina del primero al segundo centenario, Tomo I*, edited by Susana Torrado. Buenos Aires: Edhasa.

———. 2010. "La historia de la estadística en las universidades y en los institutos nacionales de estadística. El caso argentine." In *Estatísticas nas Américas: por uma agenda de estudos históricos comparados*, edited by Nelson de Castro Senra and Alexandre de Paiva Rio Camargo, eds. 2010. Rio de Janeiro: IBGE. https://biblioteca.ibge.gov.br/visualizacao/livros/liv44323.pdf. Accessed 28 September 2020.

———. 2013. "Representaciones estadísticas de la vejez. Argentina, 1869–1947." *Revista Latinoamericana de Población* 7 (13): 5–28. https://doi.org/10.31406/relap2013.v7.i2.n13.1.

———. 2018. "Socio-History of Statistics on Latin America: A Review." *Histoire & Mesure* 33 (2): 13–32. https://doi.org/10.4000/histoiremesure.7934.
Pantaleón, Jorge. 2009. *Una nación a medida*. La Plata: Ediciones Al Margen. https://www.ides.org.ar/sites/default/files/pub/laotraventana-2009-pantaleon.pdf. Accessed 28 September 2020.
Paschel, Tianna S. 2013. "'The Beautiful Faces of my Black People': Race, Ethnicity and the Politics of Colombia's 2005 Census." *Ethnic and Racial Studies* 36 (10): 1544–1563. https://doi.org/10.1080/01419870.2013.791398.
Peachy, Dominic Keith. 1976. "The Revillagigedo Census of Mexico 1790–1794: A Background Study." *Bulletin of the Society for Latin American Studies* 25: 63–80. https://www.jstor.org/stable/45179625. Accessed 28 September 2020.
Pearce, Adrian J. 2001. "The Peruvian Population Census of 1725–1740." *Latin American Research Review* 36 (3): 69–104. https://www.jstor.org/stable/2692121. Accessed 28 September 2020.
Peloso, Vincent, and José Ragas. 2001. "Estadísticas y sociedad en el Perú poscolonial: el desconocido censo de Lima de 1860." *Histórica* 25 (2): 275–293. http://revistas.pucp.edu.pe/index.php/historica/article/view/8720. Accessed 28 September 2020.
Pereyra, Diego Ezequiel, and Hernán González Bollo. 2020. "Tres perfiles de funcionarios estadísticos del Estado argentino bajo el peronismo clásico (1944–1955)." *Quinto Sol* 24 (2): 1–21. https://cerac.unlpam.edu.ar/index.php/quintosol/article/view/3575. Accessed 28 September 2020.
Pérez Eyzaguirre, José Ignacio. 2010. "Los primeros censos chilenos de población (1854–1920). Análisis crítico de las fuentes de datos censales y sugerencias de uso." *Boletín de la Academia Chilena de la Historia* 119 (1): 55–95. https://www.institutodechile.cl/wp-content/uploads/2019/02/boletin_119-1.pdf. Accessed 28 September 2020.
Pinto, Jorge. 2002–2003. "Censos e imaginarios en el Chile del siglo XX." *Dimensión Histórica de Chile* 17–18: 155–190.
———. 2009. *La población de la Araucanía en el siglo XX. Crecimiento y distribución espacial*. Temuco: Ediciones Universidad de La Frontera.
———. 2010. *Los censos chilenos del siglo XX*. Temuco / Osorno: Universidad de La Frontera / Universidad de Los Lagos.
Piza, Edith, and Fulvia Rosemberg. 1999. "Color in the Brazilian Census." In *Race in Contemporary Brazil: From Indifference to Inequality*, edited by Rebecca Reichmann. University Park: Pennsylvania State University Press.
Ponce de León, Macarena, Francisca Rengifo, and Ernesto San Martín. 2010. "¿Demanda social por educación? Estadística, sociedad y alfabetización a partir del Censo de la República de 1854." *Revista Pensamiento Educativo* 47

(2): 303–316. http://pensamientoeducativo.uc.cl/index.php/pel/article/view/472. Accessed 28 September 2020.

Prieto, Fabián. 2005. "Anatomía de la población colombiana: la técnica estadística en Colombia y el levantamiento del censo de población en 1912." *Memoria & Sociedad* 9 (19): 55–67. https://revistas.javeriana.edu.co/index.php/memoysociedad/article/view/7890. Accessed 28 September 2020.

Ragas, José. 2008. "Ideólogos del Leviatán. Estadística y sociedad en el Perú (1791–1876)." In *Intelectuales y poder. Ensayos en torno a la república de las letras en el Perú e Hispanoamérica (ss. XVI–XX)*, edited by Carlos Aguirre and Carmen McEvoy. Lima: Instituto Francés de Estudios Andinos / Instituto Riva Agüero.

Ramos Zincke, Claudio. 2015. "Poverty as Epistemic Object of Government: State Cognitive Equipment and Social Science Operations." *Social Science Information* 54 (1): 91–114. https://doi.org/10.1177/0539018414556178.

Rayes, Agustina. 2013. "'Bestias negras de la estadística'. Las exportaciones argentinas 'a órdenes' (1895–1913)." *Estatística e Sociedade* 3: 6–20. https://seer.ufrgs.br/estatisticaesociedade/article/view/39172. Accessed 23 June 2021.

Rodrigues Botelho, Tarcísio. 1999. "População e nação no Brasil do século XIX." PhD dissertation, Universidade de São Paulo. https://biblioteca.ibge.gov.br/visualizacao/livros/livp64679.pdf. Accessed 28 September 2020.

———. 2005. "Censos e construção nacional no Brasil Imperial." *Tempo Social* 17 (1): 321–341.

———. 2010. "Censos nacionais brasileiros: da estatística à demografia." In *Estatísticas nas Américas: por uma agenda de estudos históricos comparados*, edited by Nelson de Castro Senra and Alexandre de Paiva Rio Camargo. Rio de Janeiro: IBGE. https://biblioteca.ibge.gov.br/visualizacao/livros/liv44323.pdf. Accessed 28 September 2020.

Román Antequera, Alejandro. 2019. "Entre dos imperios: un elemento de control común y divergente. Las estadísticas demográficas en Cuba, Puerto Rico y Filipinas." *Amérique Latine Histoire et Mémoire. Les Cahiers ALHIM* 37. https://doi.org/10.4000/alhim.7695.

Romero Marchesini, Natalia. 2021. "Mortes que contam: a produção de números sobre feminicídios, transfeminicídios e travesticídios como uma política de estado." *Mediações - Revista de Ciências Sociais* 26 (2): 256–271. https://doi.org/10.5433/2176-6665.2021v26n2p256.

Santos, Marco Aurélio Martins. 2006. "Demógrafos-sanitaristas: vida e morte pelas estatísticas." In *História das estatísticas brasileiras. Estatísticas legalizadas: (c. 1899–c. 1936), Volume II*, edited by Nelson de Castro Senra. Rio de Janeiro: IBGE.

Santos, Júlio César Paixão. 2018. A circulação das ideias de estatística no Segundo Reinado: periodismo, discurso científico e ciências biomédicas no Rio de Janeiro (c.1840–1870). Rio de Janeiro, RJ. Tese de Doutorado. Fundação Oswaldo Cruz, 325 p.

Scarano, Francisco. 2009. "Censuses in the Transition to Modern Colonialism: Spain and the United States in Puerto Rico." In *Colonial Crucible: Empire in the Making of the Modern American State*, edited by Alfred W. McCoy and Francisco A. Scarano. Madison: University of Wisconsin Press.

Schwartzman, Simon. 1997. "Legitimidade, Controvérsias e Traduções em Estatísticas Públicas." *Teoria & Sociedade* 2: 9–38. http://www.schwartzman.org.br/simon/estpub.htm. Accessed 28 September 2020.

Senra, Nelson de Castro. 1997. "Um olhar sobre os anuários estatísticos." *Ciência da Informação* 26 (1): 7–11. http://revista.ibict.br/ciinf/article/view/747/774. Accessed 28 September 2020.

———. 2005. *O Saber e o Poder das Estatísticas*. Rio de Janeiro: IBGE. https://biblioteca.ibge.gov.br/visualizacao/livros/liv83121.pdf. Accessed 28 September 2020.

———. 2006–2009. *Historia das estatísticas brasileiras (1822–2002), Volumes 1–4*. Rio de Janeiro: IBGE. https://memoria.ibge.gov.br/publicacoes/historia-das-estatisticas-brasileiras.html. Accessed 28 September 2020.

Senra, Nelson de Castro, and Alexandre de Paiva Rio Camargo, eds. 2010. *Estatísticas nas Américas: por uma agenda de estudos históricos comparados*. Rio de Janeiro: IBGE. https://biblioteca.ibge.gov.br/visualizacao/livros/liv44323.pdf. Accessed 28 September 2020.

UN-CEPAL. 2009. *Censos 2010 y la inclusión del enfoque étnico: hacia una construcción participativa con pueblos indígenas y afrodescendientes de América Latina*. Santiago: UN-CEPAL. https://repositorio.cepal.org/bitstream/handle/11362/6940/1/S0900568_es.pdf. Accessed 28 September 2020.

Vera, Héctor. 2017. "Counting Measures: The Decimal Metric System, Metrological Census, and State Formation in Revolutionary Mexico, 1895–1940." *Histoire & Mesure* 32 (1): 121–140. https://doi.org/10.4000/histoiremesure.5780.

Villamarín, Juan, and Judith Villamarín. 1981. "Colonial Censuses and Tributary Lists of the Sabana de Bogotá Chibcha: Sources and Issues." In *Studies in Spanish American Population History*, edited by David J. Robinson. Boulder: Westview Press. https://doi.org/10.4324/9780429307805.

Wainerman, Catalina, and Zulma Recchini de Lattes. 1981. *El trabajo femenino en el banquillo de los acusados. La medición censal en América Latina*. México, D.F.: Terranova.

Williams, John Hoyt. 1976. "Observations on the Paraguayan Census of 1846." *Hispanic American Historical Review* 56 (3): 424–437. https://doi.org/10.1215/00182168-56.3.424.

The manufacturer's authorised representative in the EU is Springer Nature Customer Service Centre GmbH, Europaplatz 3, 69115 Heidelberg, Germany. If you have any concerns regarding our products, please contact ProductSafety@springernature.com

Printed and bound by CPI Group (UK) Ltd, Croydon, CR0 4YY
01/10/2026
02225895-0001